SPIDERWEBS AND SILK

SPIDERWEBS AND SILK

Tracing Evolution from Molecules to Genes to Phenotypes

CATHERINE L. CRAIG

2003

Oxford New York
Auckland Bangkok Buenos Aires Cape Town Chennai
Dar es Salaam Delhi Hong Kong Istanbul Karachi Kolkata
Kuala Lumpur Madrid Melbourne Mexico City Mumbai Nairobi
São Paulo Shanghai Taipei Tokyo Toronto

Published by Oxford University Press, Inc.,
198 Madison Avenue, New York, New York 10016

www.oup.com

Parts of chapter 7 are abstracted from: Craig, C. L., M. Hsu, D. Kaplan, and N. E. Pierce. 1999. A comparison of the composition of silk proteins produced by spiders and insects. *Int. J. Biol. Macromol.* 24: 109–118.

Library of Congress Cataloging-in-Publication Data
Craig, Catherine Lee.
Spiderwebs and silk : tracing evolution from molecules to genes to phenotypes / by Catherine L. Craig.
p. cm.
Includes bibliographical references (p.).
ISBN 0-19-512916-4
1. Orb weavers—Evolution. 2. Silk. I. Title.
QL458.42.A7 C72 2003
595.4′4—dc21 2002033394

9 8 7 6 5 4 3 2 1

Printed in the United States of America
on acid-free paper

To Robert S. Weber
and in memory of my loving parents,
Volney Howard Craig, Jr., and Allene Lee Withers Craig

Identity

1) An individual spider web
identifies a species:

an order of instinct prevails
through all accidents of circumstance,
though possibility is
high along the peripheries of
spider
webs:
you can go all
around the fringing attachments

and find
disorder ripe,
entropy rich, high levels of random,
numerous occasions of accident:

2) the possible settings
of a web are infinite:

how does
the spider keep
identity
while creating the web
in a particular place?

how and to what extent
and by what modes of chemistry
and control?

it is
wonderful
how things work: I will tell you
about it
because

it is interesting
and because whatever is
moves in weeds
and stars and spider webs
and known
is loved:
in that love,

each of us knowing it,
I love you,

for it moves within and beyond us,
sizzles in
winter grasses, darts and hangs with bumblebees
by summer windowsills:

I will show you
the underlying that takes no image to itself,
cannot be shown or said,
but weaves in and out of moons and bladderweeds,
is all and
beyond destruction
because created fully in no
particular form:

if the web were perfectly pre-set,
the spider could
never find
a perfect place to set it in: and

if the web were
perfectly adaptable,
if freedom and possibility were without limit,
the web would
lose its special identity:

the row-strung garden web
keeps order at the center
where space is freest (interesting that the freest
"medium" should
accept the firmest order)

and that
order
diminishes toward the
periphery
allowing at the points of contact
entropy equal to entropy.

— A.R. Ammons, *The Selected Poems*

Preface

Spiders have remained relatively unchanged in terms of body plan since their origin over 350 million years ago. They have maintained two body parts, eight legs, no wings, silk production, and a developmental pathway in which offspring are morphologically similar to adults. In contrast, rapid change and immense diversification of morphologies and lifestyles have characterized the evolution of their sister taxa, the insects. For instance, whereas all spiders are predators and feed almost exclusively on insects, the huge diversity of insects and mites correlates with the huge diversity of their foraging strategies—in particular, with the enormous variety of plants they feed on, animals they prey on and scavenge, and the animals they parasitize. Indeed, it is exactly their rapid morphological transformations and resulting functional diversification that is given credit for insects' unequaled evolutionary success in terms of numbers of species. Yet the relatively unchanging spiders rank seventh among all animal species in global diversity, falling behind only the five largest orders of insects (Coleoptera, Hymenoptera, Lepidoptera, Diptera, Hemiptera) and the Acari (Arachnida). They are obviously doing something right!

This something could be silk production. Silk is an Araneae trademark of sorts. Some other arthropods produce silk; in fact, silk production is present in some members of almost all insect taxa. But every spider produces silk, and only spiders produce multiple silk proteins, use them for multiple purposes, and display multiple types of silk-spinning behaviors. And while, over the course of their evolution, insects lost and later regained the ability to produce silks, silk synthesis has always been an integral component of the physiology and lifestyle of all spiders. Furthermore, tracing the evolution of silk gland systems reveals that an increased commitment over time to a silk physiology correlates with the diversification of uses to which silks are put.

Still, as successful as spiders obviously are, no evolutionary strategy comes without costs. For one thing, silk synthesis is metabolically "expensive" (as defined by the amount of ATP invested). Although all spiders use fibrous proteins or silks for different purposes, most spiders are very conservative regarding the amount of silk that they synthesize and secrete. Also, one cannot but wonder what evolutionary opportunities spiders might have missed—perhaps, for one, eusociality— by depending on a silk-based physiology. In this monograph, I explore some of the major questions regarding the evolution of spiders and in particular the Araneoidea (orb-spinners). It may be stated that the spiders' status as one-of-a-kind occupants of a lucrative lifestyle is as good as it gets—that, for all its actual and possible costs, silk production more than pays for itself. But, as in all of evolution, the current picture of spider evolution is considerably more complicated and interesting than that statement.

My interest in spiders and silk-producing systems is spurred primarily by what could be characterized as a "basic science" approach. Even so, the research results reviewed in this book may have practical applications. The primary goal of most current research on spider silks is to find ways to synthesize spider silk proteins artificially. Because spider silk is a multi-component, composite material that combines strength and toughness, it is a unique material. Unlike silk collected from silkworm cocoons, which evolved to protect developing larvae, spider silk has evolved to dissipate energy, specifically, the kinetic energy of intercepted prey. Synthetic spider silk fibers, sharing this feature, could result in the production of high-performance fabrics that can be used as protective, lightweight, water-absorbing, high-stretch textiles. Furthermore, the characteristics of synthetic silk textiles will substantially enhance garment comfort in comparison to current clothing made from nylon or Kevlar, and silk hydrophilic properties will enhance their ecological value as they can be dyed and treated with totally recyclable, natural pigments derived from plant and animal products.

Yet, although investigators have been trying to produce artificial silk proteins for almost ten years, success has been elusive. Perhaps approaching artificial silk synthesis from a new angle of understanding, an evolutionary angle, might reveal new ideas and alternative approaches to the artificial silk synthetic problems.

What silks should be studied? In fact, silks are produced by all spiders and almost all insects. Silks produced by insect larvae and spiders are made up of different types of repetitive sequences, different types of gene organization and result in different physical properties. Despite (or perhaps because of) this variation, relatively few insects have evolved the ability to "spin" or draw them into weight-bearing threads. Spinning behavior has evolved only in the most derived insects (with one exception); substrate-free spinning has evolved only in the true spiders. Contrasting the natural diversity of silk composition and organization, coupled with the natural diversity of physiological systems in which silk proteins are processed, and the diversity of spider silk-spinning behavior might redirect current experimental efforts to new fruitful approaches.

My own approach to studying this fascinating system is to use any information available on any type of silk. In particular, I have made extensive use of information on the silks produced by insects, especially by the moth larvae *Bombyx mori*, to determine whether spiders are unique, what in the realm of evolution they might have done differently (or better), and in what cases spider and insect silk systems converge. I have crossed all of the spatial and temporal scales that I can imagine. Indeed, one of my goals is to illustrate the ease with which evolutionary studies of spiders and silk proteins can cross traditional molecular and organismal borders. Like any evolutionary analysis, however, all of the inferences presented in this monograph should be viewed as a hypothesis.

What are silks?

Biologists and biochemists define silks differently. To a biologist, silks are secreted, fibrous materials that are deposited or spun by organisms. To a biochemist, silks are protein threads composed of repeating arrays of polypeptides that contain both discrete crystalline and noncrystalline domains that are oriented around a fiber axis. These two definitions—one nonspecific and one relatively precise—reflect the biologist's insight into the diversity of similar-looking secreted materials and the biochemist's insight into the structural organization of proteins. These two definitions also suggest that there is a continuum of secreted proteins that have a fibrous component and are characterized by different degrees of organization. In chapter 1, silks are broken down into their amino acid components and then reconstructed into complex protein threads. The goal of this parts-to-proteins approach is to provide a foundation for understanding why some silks take one structural configuration and other silks another. In some cases, structure seems derived from amino acids alone; in others, structure is clearly the result of processing systems. The diversity of processing systems and the amount of "evolutionary energy" or focus devoted to them seems to correlate with the importance of the silks to organismal survival. The goal of the next seven chapters is to identify the potential and realized pathways that best characterize the evolution of silk proteins and their significance to the diversity of spiders.

Silk proteins are the perfect system through which to study evolutionary conflicts among molecular genetic constraint, protein architectural constraint, protein diversity and selection. To date, silk genes have been found to be organized in two ways. Fibroin silk genes produced by some moth larvae have a two-exon, one-intron organization whereby crystalline and noncrystalline regions are completely integrated. In contrast, the flagelliform gene (spiders) and the gene for fibrous protein glues (insects and spiders) are made up of multiple exons and introns. In the fibrous protein glues, crystalline and noncrystalline regions are encoded in different exons. In flagelliform silks, all of the repeating exons are similar and no crystalline domains are present. Due to the highly repetitive nature of all fibroin silks, slippage and frequent cross-over errors are likely to occur during recombination. Some sequences are even identified as mutational hotspots because they are error-prone. Theory predicts that these "mutator sequences" can accelerate the rate of adaptive evolution in rapidly changing environments. Silks, therefore, can be described as having "high evolvability" due to their variation-generating mechanisms. However, if selection constrains silks to specific functional properties, then the effects of sequence instability to generate new proteins may be limited. Are the different gene architectures outlined above significant either to the stabilization of the gene or the adaptation of proteins to different environments? Chapter 2 outlines the patterns of gene organization of different silks to provide a better understanding of how new silks might evolve.

Chapter 3 explores how protein variation, in the context of protein organization, correlates with the evolution of silk function. At least two and possibly three major speciation events correlate with the evolution of new types of silk glands among spiders: (1) the divergence of the two major groups of Mygalomorphae (the Fornicephalae and Tuberculotae), the evolution of tubular silk glands from acinous glands and substrate-spinning; (2) the divergence of the Araneomorphae (the true spiders) from Mygalomorphae, the evolution of ampullate silk glands; and substrate-free spinning; and (3) the divergence of the Araneoidea from the Deinopoidea and the evolution of flagelliform silk glands. Each of these events correlates with a major shift in silk use and functional properties. The evolution of the

Tuberculotae correlates with the evolution of above ground foraging. Above ground, spiders extend their foraging area with fibrous proteins that adhere to each other as well as to the surface on which they are laid. While these silks contain crystallite inclusions, the crystallites are not organized or well oriented. The evolution of the araneomorphs and ampullate silks, the first silks spun into weight- bearing threads, correlate with the evolution of an irregular aerial net. Ampullate gland silks contain well-organized, well-oriented β-pleated sheets that make up their crystalline domains. The evolution of the Araneoidea and the flagelliform silk, flexible and elastic threads containing no β-pleated sheet structures, correlates with the evolution of a symmetrical, vertical net that is a spider's evolutionary answer to fast-flying prey. Once again, silks are taken into a new selective domain. Here silks must absorb prey impact, and either attract prey to the web area, or disappear against their vegetation backgrounds. Chapters 4–6 are devoted to outlining these seemingly exclusive properties of silks and their effects on spider foraging phenotype. Discussions of the flight and visual physiology of insects will serve as a foundation for the hypotheses that will be proposed and tested.

Particular web designs, for example, the density of the fibers, or the presence of viscid glue droplets, affect not only the functional properties of the web, but also its visibility to potential prey. In chapter 4 we first discuss several aspects of insect spatial vision. We then provide experimental evidence to support a hypothesis that the properties of insect visual systems might have constituted a selective factor in the evolution of silk systems. Insects are able to detect spider webs under a variety of light conditions, provided that the web produces sufficient contrast against its background. Spider silks vary in brightness, that is, in the amount of light they reflect, and will therefore be either visible or invisible to the insect, depending on the background against which they are suspended.

In the course of evolution, shifts in silk reflectance properties are correlated with shifts in the structural properties of silks and hence their mechanical functions and optical properties. The X-ray diffraction studies detailed in chapter 3 suggest that the proto-silks are protein glues and contain unoriented crystallites that scatter light in all directions. The araneomorph spiders evolved spinning behaviors that resulted in threads with highly oriented crystalline domains. In these cases, silks are reflective only from specific angles. The most recently evolved silks (flagelliform or *Flag* silk) contain reduced or no crystalline components at all. Correlated with the loss of crystalline domains is the evolution of translucence. One additional, important feature of this evolutionary step, is the evolution of viscous glues. While *Flag* silks do not reflect light, the viscous glues disperse and scatter light broadly. Thus, with the evolution of transparent silks and viscid glues, selection could act independently on the mechanical properties of silks and independently on the physical properties of the glues. The evolutionary success (here defined as diversification) of the Araneoidea might, at least in part, be due to these factors.

Silks reflect specific amounts of light and therefore they differ in brightness. They may differ, in addition, in the spectral composition of the light that they reflect, and therefore in color. When a web is viewed against a differently colored background, it will produce color contrast. As in the case of achromatic contrast (chapter 4), a sufficient amount of color contrast may render the web visible. In chapter 5 we deal with the color vision system of insects and its possible impact on the evolution of silk colors.

Insects possess three spectral types of photoreceptors that together cover a spectral region between about 300 nm and 700 nm. The absorption maxima of the photoreceptors are in the ultraviolet (UV), blue, and green regions of the spectrum. Silks that reflect similar amounts of light at all wavelengths appear white; silks that reflect light only one specific region of the light spectrum appear colored. The silks produced by the most ancestral spiders are characterized by a reflection peak in the UV region of the spectrum, thus producing contrast against backgrounds that lack UV reflection. This contrast might serve as a spectral cue to insects even in dim light conditions, because the insects' UV photoreceptors are particularly sensitive.

Because the sun and the sky are the only natural sources of UV, insects may use UV cues to determine the position of the sky. When early spiders began to produce aerial webs from UV-reflecting silks, the UV-reflection of cribellate silk may have attracted insects, particularly those that seek open sites (as in the so-called open-space response). Silks produced by the cribellate orb weavers also reflect UV, as well as additional spectral peaks in the blue and green spectral regions. In some cases, however, the color of silks might have evolved specifically to render the webs invisible against their background, causing insects to fly into them by error. The only way to test these hypotheses is through experimental studies *in situ*. Chapter 6 focuses on the effects of silk color and pattern on insect perception and interception at webs in natural conditions.

The experiments discussed in chapter 6 attempt to determine the cues that an insect sees as it approaches a spider web, or, as importantly, the cues that it does not see. Do webs intercept any insect that flies? The findings described in the previous chapters suggest that spiders attract specific types of prey through the evolutionary manipulation of silk colors and patterns. To discriminate among different possible effects, the experiments described in chapter 6 are designed to ask insects questions in a way that promises clear answers. To determine whether or not silk color alone affects the insect's response to webs under different natural conditions, stingless bees, *Trigona fulviventris* (Meliponidae; one of the common prey of *Nephila clavipes*), were trained to feed at several different sites in the forest understory (either dim light conditions—closed canopy—or bright light conditions—open canopy). Once trained to the various feeding sites, the bees were presented with natural pigmented and unpigmented webs spun by *Nephila clavipes*. The bees were also presented with webs spun by *A. argentata* that differed in fiber density but were a similar color to the unpigmented webs spun by *Nephila*. The bees' behavioral responses measured in these studies show that the details of the web's site, such as background vegetation and ambient light intensity, affect the visibility of the web significantly. The results further suggest that the foraging modes employed by spiders may have evolved in response to visual selection by their prey in the various light environments.

Some spiders spin low-visibility webs but decorate them with high-visibility silks. Both the color and pattern of the decoration are similar to the colors and patterns of resources that foraging insects seek. Hence, because of the high visibility of the decoration silks, insects—in fact both predators and prey—are drawn to the web site. A further complicating factor of this foraging mode, is that if an intercepted insect struggles free of the web, it can then use the decorative pattern to remember the web site at future encounters. In this case, learning plays an important role in insect response to webs, as well as the evolution of spider behaviors that may have evolved to disrupt insect learning and memory.

One factor that has not been discussed so far is the obvious metabolic drain of silk synthesis to the organisms that produce silks. In the case of moth larvae, silk systems produce large volumes of protein during a very limited period of larval life, but this is a period when resources are most likely to be abundant and predictably available. Spiders, in contrast, produce silks throughout their lives and, because they are predators, resource availability is likely to fluctuate. It could be that the diverse fibrous proteins and silks produced by spiders evolved under selection simply to produce a needed volume of protein when food availability is limited. In this scenario, new glands might evolve making use of alternative amino acids and alternative physiological systems to meet an energetic need. The fact that silks evolved having different physical properties could simply reflect what silk systems came up with in response to cost selection. Chapter 7 weighs the potential role of inter-gland competition for amino acids, and hence energy, in the evolution of diverse proteins.

By using the central metabolic pathways, the metabolic costs of silk production for insects (largely herbivores) and spiders (predators) are compared. Silk cost is calculated as the amount of energy sacrificed when the breakdown of glucose is diverted to an amino acid plus the ATP invested in its synthesis. Putting aside all previous hypotheses, these studies suggest that the complex silk-producing systems of spiders evolved via inter-gland competition for amino acids. Most of the amino acids that make up the silk of ancestral spiders are gathered from the environment. However, the evolution of ampullate gland silks, largely composed of amino acids that the spider synthesizes, correlates with the evolutionary speciation of the Araneomorph spiders (97% of all spider species). The most recently evolved silk, flagelliform silk, is the least costly silk spiders produce, despite the fact that its repeating modules contain the most expensive amino acid, proline. This, too, supports the idea that spider silk glands evolved via inter-gland competition because the flagelliform silk gland evolved making use of an amino acid largely absent in all other silks. The fact that the last gland evolved makes use of the most costly amino acid that spiders synthesize, and the fact that when a new gland evolves, the ancestral glands are retained, suggests that the evolution of the complex silk-producing systems of spiders could be reduced to a problem of energetics.

Spiders have confronted the same evolutionary challenges as insects by making creative use of silk proteins. For example, adult insects fly to disperse, but juvenile spiders disperse on airborne silk threads. Predatory insects catch prey on the wing; spiders spin webs that intercept them. There is one aspect of insect lifestyle, however, that spiders have not been able to match. That is the evolution of highly eusocial societies with allocated reproductive function and castes of morphologically diverse individuals.

The primary advantage of eusociality is energetic. It is hard to imagine why spiders would not benefit from the evolution of a eusocial system, particularly in light of the fact that most spiders are not able to recycle the silks they spin. Imagine what a eusocial spider society might look like. Spider reproductives would lose their silk gland system to maximize the energy they devote to egg production. Worker spiders would lose reproductive function while maximizing the energy they devote to silk synthesis to extend the web, catch prey, and feed the queen spider. Spider soldiers would loose both reproductive and silk systems while evolving costly venoms to stun predators and prey. Why has this system not evolved? After all, spiders have been around for at least as long as the insects and share an evolutionary ancestor.

Multiple selective factors have been proposed that favor the evolution of eusociality in insects. These include a defensible nest or food site, the potential for extended periods of development, monogamous reproductive pairs, high chromosome number (hence reduced variance in shared genes), inheritance of resources, and cooperative defense against predators, parasites, and competitors All of these factors also affect spiders—except one. That exception is the potential for extended periods of development. To date, only one regulatory hormone, ecdysone, has been identified in spiders. The physiological ability to modify the timing of developmental events in the insects, however, is due to the interacting effects of a second hormone, juvenile hormone. Is there a conflict between the physiology that enables the life-long production of silks and the physiology that enables higher eusociality? Alternatively, are complex regulatory hormones the only pathway through which reproductive and worker castes can evolve?

Studying the evolution of silk production systems is both scientifically rewarding and fascinating. All aspects of this biological system, from genes to protein structure to function to the ecological context in which the system is evolving are accessible, measurable, and can be manipulated. The problems are diverse and it is possible to study their evolutionary interaction from many perspectives. Therefore, if one avenue of investigation is stymied, perhaps due to the lack of a technological advance, there is always another direction from which the same or a similar question can be approached that gives a slightly different but complementary insight. The status of "evolutionary success" can clearly be ascribed to spiders because of their numerical dominance as well as their status as the most important predators of insects, and their unique status as the only organisms that have evolved to make their living through a complex, silk-producing system.

My goal in writing this monograph has been to explore spiders and an area of evolution that has fascinated me for the past twenty years, and to present a range of evolutionary questions, some new and some old, that can be investigated through studies of spider and silk systems. Perhaps the most important of these has been to describe a system and methods where a link between organismal ecology, gene, and protein evolution can be quantified and linked to rates of evolutionary change. This is, of course, just an opinion. Other investigators will hopefully be stimulated to explore more deeply the many other evolutionary issues discussed in this book ranging from molecules to genes to proteins to organisms.

Acknowledgments

I would like to acknowledge the many friends and colleagues who supported me while writing this book. I am particularly grateful to Robert May and Simon Levin for encouraging me to embark on this project, to Naomi Pierce for welcoming me to Harvard where much of the writing has taken place, and to Gonzalo Giribet for providing me working space in the Museum of Comparative Zoology. I also am grateful to Francie Chew and David Kaplan, who introduced me to the Tufts faculty and students and the unique set of resources that university has afforded me.

Many students and collaborators assisted with the research on which this book is based. In particular, Robert Weber, Gary Bernard, Viggo Andreasen, Rudolfo Contreras, Mary Derr, Naomi Donnelly, Kathy Ebert, Corey Freeman, Mark Hauber, Cheryl Hayashi, Rachael Hinden, Michael Hsu, Nicole Herron Kolhoff, Paula Kovoor, Jenn Maas, Gayle Matthews, Nzomi Nishimura, Karen Nutt, Tim Schaffer, Viviany Taqueti, Joseph West, Lisa Reed, and Eve Vogel each enhanced this work with their unique ideas and hard work. Many colleagues sent silk from around the world: Jon Coddington, Fred Coyle, Janice Edgerly, Charles Griswold, Marie E. Herberstein, Brent Opell, Jason Bond, and Jackie Palmer. In addition, I was particularly honored to have had the opportunity to work with a dear colleague, Akira Okubo, who was always encouraging, always insightful, and always a friend.

All of my field studies were completed at the Smithsonian Tropical Research Institute, Panama, and my debt to the staff at STRI, as well as to RENARE for permission to collect and export spiders and stingless bees, is immeasurable. SCM Chemicals supplied needed pigments and Li-Cor assisted with light measurements. The financial and moral support of the John Simon Guggenheim Foundation allowed me to focus on the book full time, and the Mary Ingrahm Bunting Institute provided two years of fellowship and a unique environment in which to work. The staff of the Ernst Mayr Library, in particular Mary Sears, ensured that every reference I needed was at my disposal. Diana Wheeler, Brent Opell, and Jason Bond read portions of the manuscript, suggested revisions and references I had missed. Miriam Lehrer started out as a reviewer and made so many excellent comments that she became an author. The X-ray data on protein structure presented in chapter 3 is original work done with Christian Riekel. I am grateful that he agreed to coauthor that chapter with me. We both thank D. Sapede for contributions to the X-ray data analysis, M. Burghammer for help with X-ray data collection, I. Sinigireva for SEM images, and Carl Brändon for interesting and educational discussions. Andrew Berry and Daniel Weinreich, too, were particularly generous with their time, discussions, and thoughts. I will always be grateful that Oxford University Press and Kirk Jensen allowed me to work on this book without imposing deadlines that I could not meet.

Funding for the research presented here was provided by the American Philosophical Society, A.P. Sloan Foundation, the Office of Naval Research, National Geographic Society, National Science Foundation, Whitehall Foundation, the Smithsonian Institution, Radcliffe Research Partners Program, and New England Consortium for Undergraduate Education (NECUSE). Finally, I wish to express my thanks to my editor, Leslie Brunetta, for her keen eye and help in expressing many of the ideas in this book.

Contents

SPIDERWEBS AND SILK

1

Silk Proteins: Breakdown and Evolutionary Pathways

To understand the evolution of spiders, one must understand their silk-producing systems. This chapter will review our current knowledge concerning the molecular composition and structure of silks and fibrous protein glues. The hypotheses generated here are based on comparison between the silks and protein glues produced by spiders and the silks and protein glues produced by insects. These comparisons, in turn, depend on understanding the molecular composition and structure of silks and glues.

Silks and fibrous proteins are made up of amino acids that exhibit diverse secondary and tertiary configurations

Only animals in the classes Insecta and Arachnida (Phylum Arthropoda) produce silks. Silks are fibrous proteins made up of repetitive sequences of amino acids. These amino acids are stored in the animal as a liquid and configure into fibers when sheared or "spun" at secretion. Animals use silks for a variety of purposes: to construct protective shelter; to provide structural support (for example, for eggs and egg sacs); and in reproduction, foraging, and dispersal (table 1.1). Arthropods also produce protein glues, which they generally use in combination with other materials, including

Table 1.1 Comparison of ways that insects and spiders use silks.

Function	Example for insects	Example for spiders
Protective shelter	Cocoon silks produced by Lepidoptera (Stehr 1987)	Retreat silks produced by all spiders (Nentwig and Heimer 1987)
Structural support	Silk egg stalk produced by Neuroptera (Stehr 1987)	Egg sac suspension threads produced by Araneidae (Foelix 1996)
Reproduction	Restriction of female movement during mating by Thysanoptera (Smith and Watson 1991)	Sperm web produced by all male spiders (Kovoor 1987)
Foraging	Under water silk prey capture nets produced by Trichoptera (Stehr 1987)	Aerial nets produced by Araneidae (Foelix 1996)
Dispersal	Dispersal of new hatched Lepidoptera (Stehr 1987)	Dispersal of newly hatched spiderlings (Greenstone 1982)

silks, to construct shelters, feeding tubes, protective enclosures, and other structures. The distinction between protein glues and silks is not, in fact, always so clear-cut, except at the extremes of definition. While others may define silks and fibrous proteins differently, in this book I am defining fibrous proteins as proteins that are deposited, and silks as proteins that are spun or pulled from the animal under tension. In general, protein glues seem to be less organized at the molecular level than spun silks.

The molecular backbone of all proteins consists of a chain of amino acids (figure 1.1), each of which is built of four components. Three of the components, an amine group (-NH_2), a carboxyl group (-COOH), and a hydrogen group (-H), are common to all amino acids and are bound to a carbon molecule designated as the α-carbon. The backbone of the protein forms when a terminal carboxyl group of one amino acid binds with a terminal amine group of an adjacent amino acid, eliminating water to make an amide bond. The fourth component of each amino acid is a variable side chain. The diversity of proteins derives

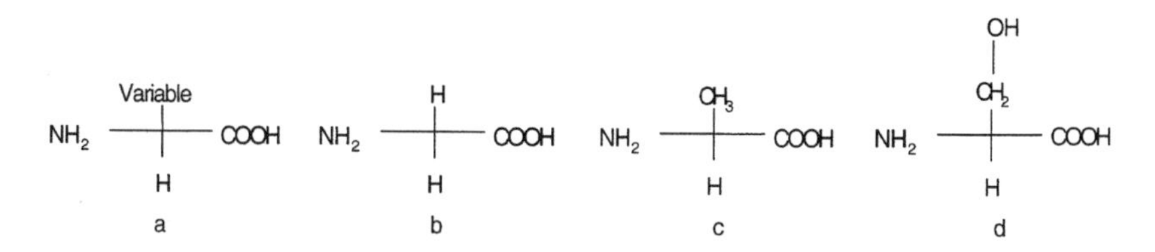

Figure 1.1 All amino acids (a) are composed of an amine group, a carboxyl group, a hydrogen group, and a side chain that is variable. The most common amino acids that make up silks are (b) glycine, (c) alanine, and (d) serine.

from the different size, shape, charge, hydrogen-bonding capacity, and chemical reactivity of these distinguishing groups. Furthermore, rotation around the α-carbon of the amino acid affects the interactions among the side chains (determines the polarity of the molecule) and allows the protein to fold in a variety of ways (figure 1.2).

Twenty different R groups give rise to 20 different amino acids, and from these all proteins are built. However, just three amino acids dominate the polypeptide chains that make up silks and fibrous proteins: glycine, alanine, and serine. Glycine (figure 1.1b), the simplest of all amino acids, has just one hydrogen atom (-H) as its side chain. Alanine (figure 1.1c), the next simplest amino acid, has a single methyl group ($-CH_3$) as its side chain. Serine (figure 1.1d) is similar to alanine, except that one of the hydrogen atoms in the side chain is replaced by a hydroxyl group so that the side chain becomes $-CH_2OH$. The proportion of fibrous proteins and silks made up of these three amino acids is variable. Other amino acids, some of which the animal produces and some of which are gathered from the environment, make up the remainder of the protein sequences. (Lucas et al. 1960).

Fibrous proteins and spun silks contain multiple protein domains that can include α-helical structures and different β-forms as well amorphous domains (figure 1.3). The crystalline, parallel-β pleated sheet structure, however, can be induced from any of these config-

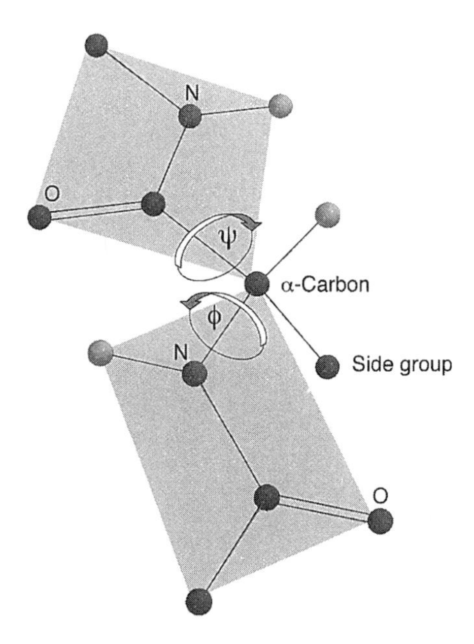

Figure 1.2 Protein folding patterns are affected by the rotation of the amide group around the α-carbon of the amino acid. Free movements are only possible around the C_α–N bond (θ) and the C_α–C bond. (After Geis in Voet and Voet 1995.)

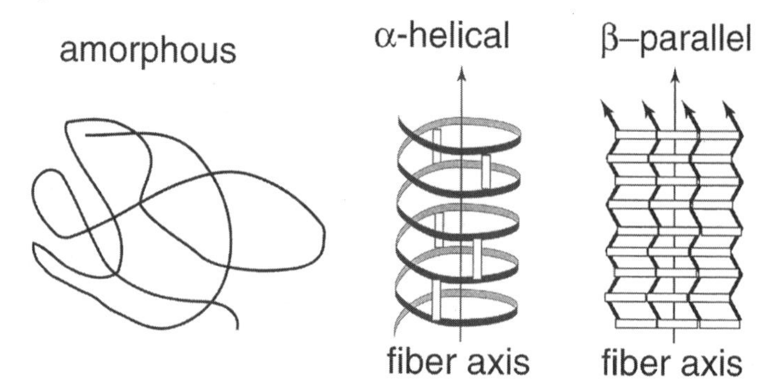

Figure 1.3 Multi-domain proteins are composed of two or more types of structural units. These include "amorphous" proteins, α-helical coils whose structure is maintained by bonds within the protein chain and parallel-β pleated sheet proteins maintained by bonds among protein chains.

urations given the appropriate conditions (figure 1.4). Those proteins characterized as coiled or α-helical are simply secreted or deposited. Both the coil and α-helical form, however, can be transposed into proteins with β-forms or oriented β-forms under mechanical shearing (Magoshi et al. 1994). These differences in tertiary structure suggest that an important link between the evolution of silks in primitive and derived forms lies not only in the amino acid sequence of the protein but also in the behavior by which the silk is processed, manipulated, or spun at secretion (Craig 1997; Riekel and Vollrath 2001; Vollrath and Knight 2001).

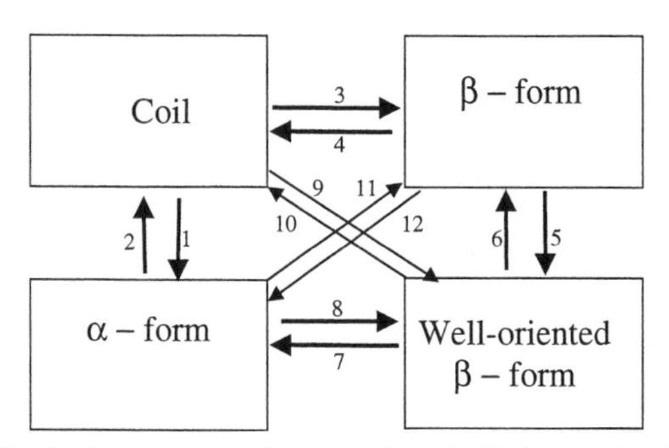

1. Crystallization from aqueous solution at pH above 6.0
2. Application of high pressure
3. Mechanical shearing
4. Application of high pressure
5. Drawing of liquid fibroin or film
6. Application high pressure
   Treatment with salt
7. Mechanical shearing
8. Treatment with a salt solution
9. Mechanical shearing; epitaxial growth
10. Application of high pressure
11. Mechanical shearing
12. Heat treatment by wet process

Figure 1.4 Conditions for crystallization of fibrous proteins accessible to biological systems. (Modified from Magoshi et al. 1994.)

Insects, as a group, produce many different types of silks and fibrous proteins; however, each individual produces only one silk protein. Spiders also produce a variety of silks and fibrous proteins but, in contrast to insects, an individual spider may produce as many as nine different types of silks and fibrous proteins, each of which may be composed of more than one type of protein (Kovoor 1987; Haupt and Kovoor 1993). Spiders produce silks with mechanical properties suitable for a wide variety of ecological functions by spinning threads with a sheath-core structure (Stubbs et al. 1992), varying the proportions of the sheath and core (Kovoor 1987), combining fibrils with different tensile strengths and extensibilities (Stubbs et al. 1992), and producing composite materials (Gosline and Demont 1984). The diverse nature of the types of fibrous proteins and silk-producing glands of spiders suggest that silk fibroins may have evolved under selection, although the specific selective events are complex and entangled. Furthermore, the presence of diverse types of glands in each spider seems to suggest that the silks and fibrous proteins produced by spiders evolved through more than one pathway.

Current hypotheses suggest that fibrous proteins produced by the Chelicerata and Hexapoda evolved independently

The most recent hypothesis of arthropod evolution suggests that the Pycnogonida are the sister taxa of the monophyletic clade Arthropoda. The Arthropoda includes the Chelicerata, the most ancestral arthropod group; the Myriapods; and the Pancrustecea, a newly named clade including the Crustacea and the Insecta (Giribet et al. 2001; figure 1.5).

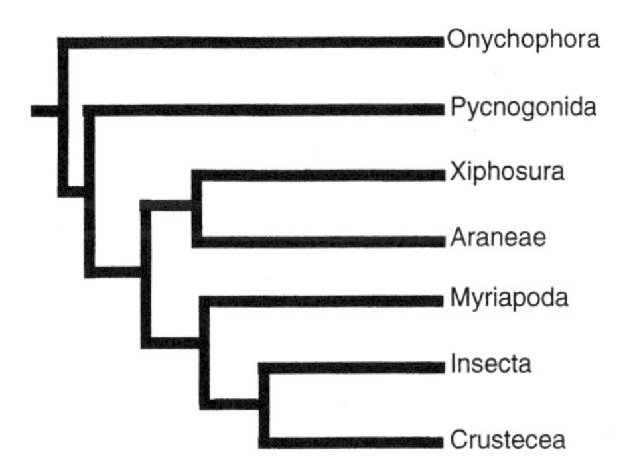

Figure 1.5 Current phylogeny of the Arthropoda and sister taxa. Jackknife values less than 50 for all clades except Onychophora-Pycnogonida (77), Pycnogonida (100) and Chelicerata (85; here represented as Xiphosura and Araneae clade). (Giribet, pers. commun.)

Systematic analyses of silk producers across the Arthropoda suggest that the glands in which silks and fibrous proteins are produced, and the types of silks and fibrous proteins produced, evolved independently and sometimes more than once (figures 1.5 and 1.6). The production of secreted fibrous proteins in the Arachnida (Chelicerata: Acari, Pseudoscorpions, Araneae), however, seems to have evolved *de novo* at least three times. Silk production is an ancestral, identifying feature of the Araneae, and almost all spiders retain both ancestral and more recently evolved silk glands. The fact that different silk types evolved at different times in spider evolution suggests that gland evolution is correlated with selection for an additional, not alternative, silk functions or metabolic needs.

Secreted fibrous proteins are present in some of the Myriapoda. Male myriapods secrete fibrous proteins from their accessory glands that they use to produce sperm webs, sperm stalks, and mating threads. Females in the Diplopoda and Symphyla also produce fibrous proteins they use for molting, egg cocoons, defense, and communication. Pseudoscorpions produce fibrous proteins that they use to enclose brood chambers (J. Adis, pers. commun.; Adis and Mahnert 1985). Regrettably, none of the macromolecular configurations of any of the fibrous proteins produced by the myriapods and pseudoscorpions have been identified (Palmer 1990). Spiders secrete fibrous proteins and silks throughout their lives. Fibrous proteins are

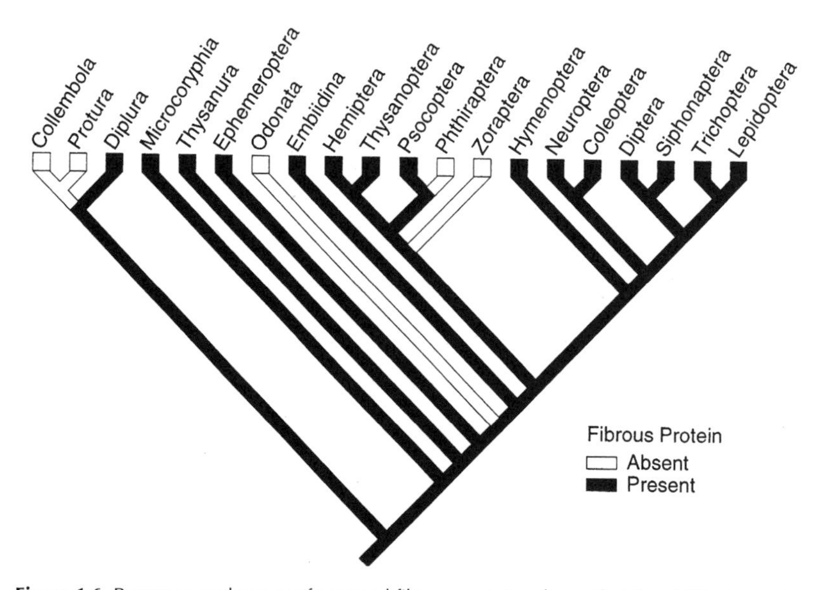

Figure 1.6 Presence or absence of secreted fibrous proteins shows that the ability to produce them has evolved multiple times across hexapod taxa. Insect phyligeny redrawn from Kristensen 1981; statistical analyses not available. (Reprinted with permission of *Annual Review of Entomology,* Volume 42. © 1997 by Annual Reviews www.AnnualReviews.org.)

secreted by almost all of the Insecta at some time during their life cycle. Silk production is not present in the Crustacea, but Crustacea secrete collagen proteins (Voet and Voet 1995).

Alpha-helical fibrous proteins produced by insects are used primarily with other materials

Fibrous proteins composed entirely of proteins in an α-helical conformation are produced not by a dedicated silk gland but by glands, such as the colleterial glands, that assume some other primary function. The primary unit of an α-helical protein is a polypeptide chain that assumes the conformation of a rod-like coil stabilized by hydrogen bonds that form between the amine and carboxyl groups within the backbone (figure 1.3). The amino acid side chains extending away from the helical array are further stabilized by H-bonds that form between polypeptide chains within protein sheets (Stryer 1995). The α-helical fibrous proteins have a relatively low glycine content but are high in acidic residues such as aspartic acid and glutamic acid (Rudall and Kenchington 1971).

Alpha-helical proteins are produced by an unrelated array of insects in the orders Dictyoptera, Siphonaptera, and Hymenoptera (Lucas et al. 1960; Warwicker 1960; Rudall and Kenchington 1971). For example, adult honey bees (Hymenoptera) produce α-helical fibrous proteins in their colleterial glands (accessory to insect genital glands) and use them to reinforce their wax combs (Hepburn et al. 1979; Hepburn and Kurstjens 1988). Larval fleas (Siphonaptera) produce α-helical fibrous proteins in their labial glands and use them to construct nests, pupation cases, and individual and group cocoons (Lucas et al. 1960; Warwicker 1960; Rudall and Kenchington 1971; Stehr 1987). Mantids use this kind of silk to construct a parchment-like ootheca (Richards and Davies 1977).

Insects producing α-helical fibrous proteins appear to use them in combination with other materials, and sometimes with other silks, suggesting that α-helical proteins provide only some of the material properties needed by the structures into which they are incorporated. The α-helical fibrous proteins of Aphaniptera, Siphonaptera, and Aculeata of the Hymenoptera (Lucas et al. 1960; Warwicker 1960; Rudall and Kenchington 1971) are characterized by reduced glycine and a high proportion of glutamic acid residues.

Beta structures are organized into pleated sheets, spirals, and helices

A second type of protein structure, and the structure that characterizes the majority of silk proteins described, is the β-pleated sheet (figure 1.3). The stacked, β-pleated sheet motif is a common structural motif in the fibroin silks produced by both insects and spiders. Beta structures differ from the α-helical structures in that the polypeptides are stabilized by H-bonds that form between amino acid chains, not within chains (Stryer 1995). The polypeptide chains that compose the β-pleated sheets are aligned side by side in a parallel or anti-parallel direction.

The β-pleated sheet organization for proteins was first observed by Marsh, Corey, and Pauling in cocoon silk produced by the silk moth *Bombyx mori* (Lepidoptera; Bombyciidae; Marsh et al. 1955b). The silks of this species assume a polar-antiparallel structure (figure 1.7). The protein sheets of these silks stack closely (intersheet distances are 3.5 Å for glycine-glycine interactions and 5.7 Å for alanine-alanine interactions) and in a regular orientation (Lucas et al. 1960; Warwicker 1960; Kodrík 1992). In contrast, when the protein chains composing β-pleated sheets run in parallel, the hydrogen bonds between the sheets are distorted. In this type of silk, the spacing between all protein sheets is about 5.27 Å. The internal strain generated by parallel packing may destabilize such silks. In fact, paral-

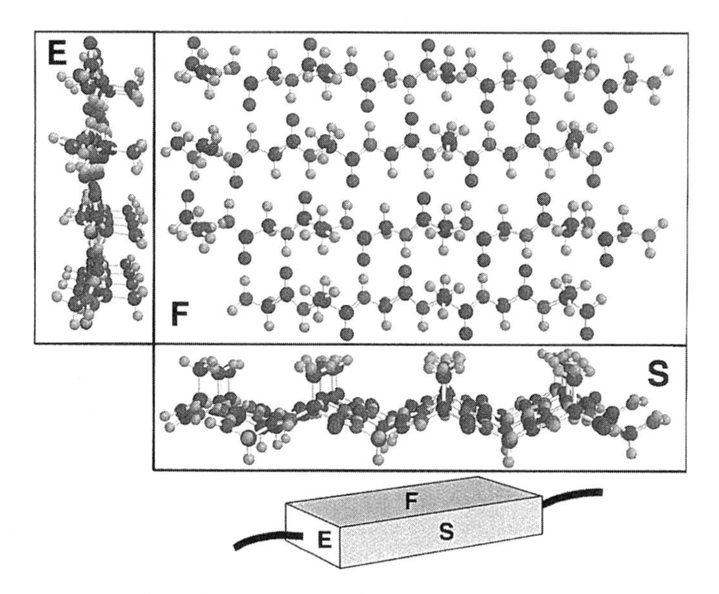

Figure 1.7 Three views, edge (E), face (F), and side (S), of a β-pleated sheet protein. The highly oriented regions of silk proteins are made up of stacked sheets of protein chains that cross-polymerize into sheets. The regular organization of the protein chains suggest the organization of a crystal and hence are called "crystalline." The details of the crystalline organization are determined by the specific amino acids that make up the repeating, polypeptide chains as well as their relative orientations. Here, the alignment of the polypeptides is anti-parallel and can be viewed from the edge, front and side and relative to the long axis of a thread.

lel-β sheets of fewer than five polypeptide chains are rare, possibly because of this distortion.

Thinking about silk protein structure has changed over the last forty years. The β-pleated sheet proteins, described as the "classical silk proteins," were originally characterized as having the polar antiparallel organization described above. Warwicker first classified them into three different groups on the basis of the structure of the silk crystallites (Warwicker 1960). The molecular symmetries of silks in Warwicker's three groups differ by increased spacing between the peptides of adjacent chains (table 1.2), which may result from increased amounts of alanine (Rudall and Kenchington 1971) and decreased amounts of glycine. The three types of "classical" silk proteins produced by Lepidoptera include:

1. The cocoon silks spun by *Bombyx mori* (Bombycoidea: Bombycidae), characterized by an intersheet packing distance of 9.3 Å and polypeptide strands of alternating amino acids of glycine and alanine or serine (Lucas et al. 1960).
2. The cocoon silks spun by the caterpillar *Anaphe moloneyi* (Bombycoidea: Thamoetopoeidae), characterized by an intersheet packing distance of 10 Å, and polypeptide strands whose repeating unit is $(\text{Gly-Ala})_x\text{-Ala})_n$ (Lucas et al. 1960; Lucas and Rudall 1968).
3. The cocoon silks spun by the caterpillar *Antheraea mylitta* (Bombycoidea: Saturniidae), characterized by an intersheet packing of 10.6 Å and polypeptides composed primarily of alanine. The dragline silk (ampullate gland) produced by the spider *Nephila madagascariensis* is also characterized by this configuration (Lucas et al. 1960).

The additional β-pleated silks that Warwicker classified into Groups 4 and 5 were then considered atypical. Group 4 silks, illustrated by the cocoon silks spun by *Thaumetopoea pityocampa* (Bombycoidea: Thaumetopoeidae) are characterized by an intersheet packing distance of 15 Å and contain large amounts of the amino acids serine and glycine (Lucas et al. 1960). Group 5 silks are characterized by an intersheet packing distance of 15.7 Å (Warwicker 1960; Lucas and Rudall 1967) and are illustrated by the cocoon silk produced by *Nephila senegalensis* (cylindrical gland silk; Araneae: Araneidae) (Warwicker 1960) as well as by a silk of unidentified origin spun by the spider *Araneus diadematus* (Araneae: Araneidae). Finally, Group 6 silks, spun by the sawflies in the families Perigidae, Argidae, and Tenthredinidae, are characterized by a parallel β-structure, but the pleated sheets are arranged into repeating units of three and include a well-defined α-helical component (Lucas and Rudall 1967; Lucas and Rudall 1968; Rudall and Kenchington 1971).

Table 1.2 Types of proteins secreted, expelled, or spun by insects and spiders.

Protein type	Protein characteristics	Glandular origin
Random coil proteins (Anderson 1970; Voet and Voet 1995)	High content of basic amino acids Randomly coiled before and after secretion Reconfigures into parallel-β proteins under stress	Insects—unknown Spiders—suspected piriform, aggregate
α-Helical structure (Lucas et al. 1960; Warwicker 1960; Lucas and Rudall 1968; Rudall and Kenchington 1971)	Low glycine, high in acidic residues Helical coil stabilized by H bonds forming between polypeptide chains Reconfigures into parallel-β protein under stress Low tensile strength, high elasticity	Insects—colleterial glands Spiders—unidentified
Parallel-β pleated sheet (Warwicker 1956; Warwicker 1960; Lucas and Rudall 1967; Rudall and Kenchington 1971)	Five primary groups characterized by increasing length of side chain and decreasing glycine Groups 1–3: Intersheet distance 9.3–10.6 Å; high tensile strength Groups 4–5: Intersheet distance 15–15.7 Å; high tensile strength and elasticity Group 6: Parallel-β sheets grouped in threes and including helical component. Mechanical properties not measured	Insects—labial glands Spiders—ampullate, flagelliform, tubular pseudoflagelliform, paracribellar
Cross-β structure (Lucas and Rudall 1968)	45% serine Polypeptide chain oriented perpendicular to fiber axis Reconfigures to parallel-β chain when stretched by a factor of 6	Insects—Malpighian tubules, adults and larvae; peritrophic membrane Unidentified in spiders

Table 1.2 continued

Protein type	Protein characteristics	Glandular origin
Collagen (Lucas and Rudall 1968; Rudall and Kenchington 1971)	30% glycine; 10% proline Three polypeptide chains twisted into triple helix stabilized by steric repulsion Tightly packed structure characterized by high strength	Insects—salivary glands
Chitin (Rudall and Kenchington 1971)	A nitrogenous polysaccharide $(C_8H_{13}NO_5)_n$ Appears to be silk, but is actually composed of cells	Insects—procuticle Spiders—unknown
Cuticulin silk (Rudall and Kenchington 1971)	Highly oriented fibrous protein of α-helical orientation	
Polyglycine II (Rudall and Kenchington 1971)	20% more glycine than Group 1 silks	

Takahashi (1994), more recently, proposed a model that allows for a greater diversity of silk molecular structure and thus differs from the classical silk models proposed in the 1960s and 1970s. He identified the alternative molecular conformation for the hydrogen-bonded, β-pleated sheet proteins in which the alanine residues of the polypeptides alternately point to opposite sides of the sheet. As a result, the crystalline region of the silk is composed of irregularly stacked, antipolar-antiparallel sheets with different orientations (Takahashi 1994). This more recent model may help in understanding the diversity of silk protein structures, in particular the "new" silks, or those spun by the derived, aerial web-spinning spiders. It may be that the conformational shifts, such as the reorientation of the amino side groups described for Group 4 and 5 structures, are only possible if silks have an anti-polar antiparallel structure as in the model proposed by Takahashi (1994).

A subset of the parallel fibrous proteins is the cross-β configuration. Polypeptide strands are oriented perpendicular to the fiber axis, instead of parallel to it (Geddes et al. 1968). Although thought to be common among arthropod silks and silk-like materials (Kenchington 1983), relatively few cross-β silks have been identified. All of these are produced by Endopterygote, insect larvae in the orders Neuroptera (*Chrysopa*; Lucas and Rudall 1968; Hepburn et al. 1979), Coleoptera (*Hydrophilus*, *Hypera*; Kenchington 1983; Rudall 1962), and Diptera (*Arachnocampa*; Rudall 1962; Lucas and Rudall

1967), and all are produced either in larval Malpighian tubules or are derived from the peritrophic membrane (figure 1.8). In one case, the lacewing *Chrysopa flava* (Chrysopidae), cross-β proteins are produced in the colleterial glands of the adult and used to attach eggs to the underside of leaves (Lucas and Rudall 1968).

Upon stretching, cross-β proteins reconfigure into a parallel-β structure similar to that of the Group 3 fibroins defined by Warwicker (Lucas and Rudall 1968). In some cases, mechanical reconfiguration of the silk may actually serve a function. It is known that the New Zealand glow-worm, *Arachnocampa luminosa* (Mycetophiloidea), for instance, intercepts flying midges (Richards 1960) with silks that exhibit a cross-β conformation (Rudall 1962; Lucas and Rudall 1967). The "unfolding" and extension of the cross-β silks may allow them to better absorb insect impact via fiber extension. The weevil *Hypera* spp. (Curculionidae: Coleoptera) produces silks that can be stretched beyond six times their original length and also results in a partial molecular transformation of the silk into a parallel-β form. The degree of shift is proportional to the degree of fiber extension (Kenchington 1983). Natural selection for silks that exhibit high extensibility may have exploited the cross-β to parallel-β transition in lieu of a gross change in the molecular composition of the silk protein. The resulting reconfiguration of the protein to a parallel-β form would enhance fiber breaking resistance. Hence, cross-β silk spinners may be able to produce silks with unique functional properties without a substantial molecular genetic change.

Comparative phylogenetic analyses pinpoint the taxa most likely to yield insight into the origins and biology of silk-producing systems

One way to explore silk evolution is to map the molecular structure of silks and fibrous proteins, the glands from which they are derived, and their functions onto a recent phylogeny of insects. However, a comparable analysis of the molecular structure of spider silks is not possible for two reasons. First, although information on the structural properties of silks and fibrous proteins produced by over 100 insect species is in the literature, only a handful of comparable data have been gathered on spider silks and fibrous proteins, and much of it is not yet published (but see chapter 3). Second, detailed studies of spider spinnerets and spigots have been used to draw the most recent phylogenies of spider taxa (Scharff and Coddington 1997; Griswold et al. 1998). The importance of the morphology of the spider's spinning organs to silk configuration is not currently understood, but is probably important. Thus, current phylogenies that are based on spinning organs are unlikely candidates for an independent test of the evolution of silk structure.

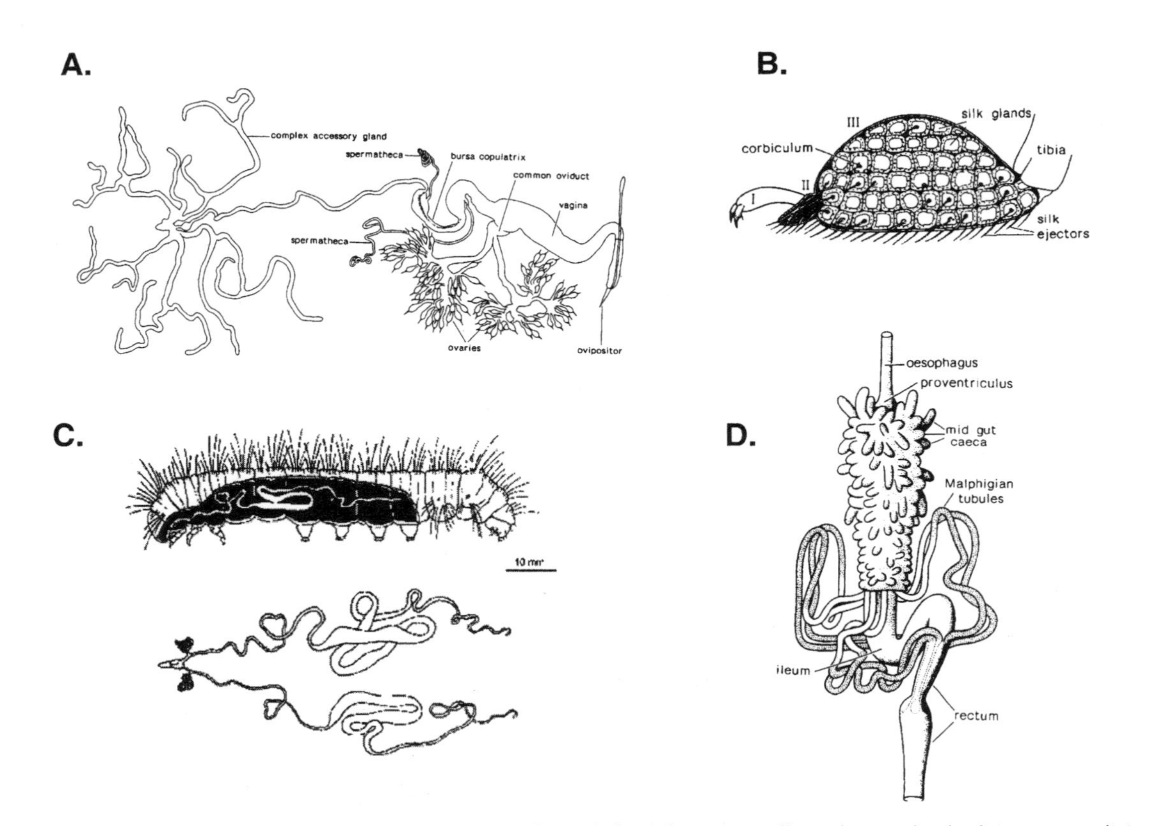

Figure 1.8 The types of insect glands in which silks are produced. (A) Colleterial gland (here a nonsilk-producing gland) of *Agrypnus caliginosus* Elateridae (from Lawrence and Britton 1991). (B) Dermal silk-producing gland of Embiidina (redrawn from Ross 1991). (C) Salivary silk-producing gland of Lepidoptera (redrawn from Fitzgerald 1995). (D) Malpighian tubules (here a nonsilk-producing gland) of *Atomaria* sp., Cryptophagidae (from Lawrence and Britton after M.E.G. Evans).

The cladograms in figure 1.6 and figures 1.9–1.11 represent the most recent hypothesis of the higher-order relationships among the Hexapoda (subsequently broadly referred to as insects; Kristensen 1981). Because the phylogenetic relationships among the insect orders Orthoptera, Dermaptera, Plecoptera, and Dictyoptera are unresolved, they were excluded from the final analysis. In addition, the analysis also excluded the insect orders Mecoptera, Raphidioptera, Stresiptera, and Megaloptera, for which no data could be found. The data available on silks and fibrous proteins and insect phylogenetics are extremely incomplete and, more often than not, the most ancestral insect type in a specific order is unknown. Furthermore, although both ancestral and derived insects secrete fibrous proteins, X-ray crystallographic data are available only for insects in the neuropteroid orders. In cases where the structure of the protein was unknown, the protein was coded simply as "fibrous" (Craig 1997). Therefore, due to the broad generalizations on which the analysis is based, the findings presented here should be used only as an indicator of which insect orders warrant more detailed study.

MacClade 3.05 (Maddison and Maddison 1992) was used to infer the "ancestral character states" for silk structure, function, and gland of origin, and MacClade's Equivocal Cycling procedure was used to determine the most parsimonious reconstruction for each. Some of

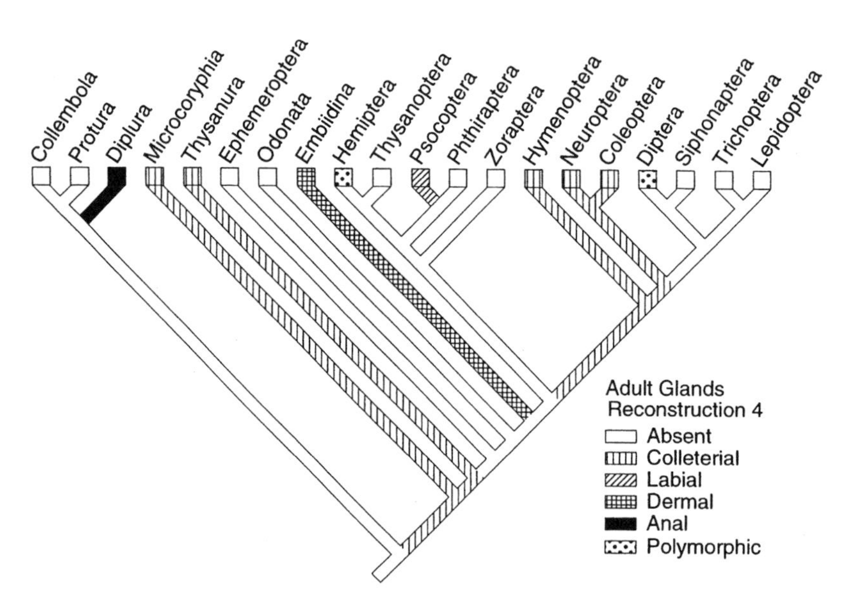

Figure 1.9 Silks produced by adult hexapods (twelve reconstructions). Silk production among adult hexapods is less common than among larvae, and all are secreted from glands that evolved for some other primary purpose. Most frequently, silks are produced by adults in colleterial glands. (Reprinted with permission of *Annual Review of Entomology*, Volume 42. © 1997 by Annual Reviews www.AnnualReviews.org.)

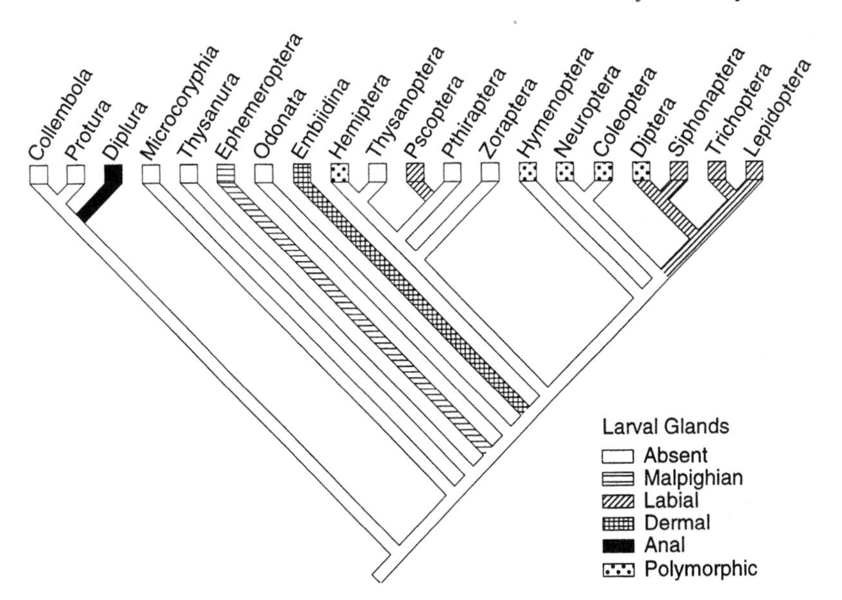

Figure 1.10 Silks produced by hexapod larvae (one reconstruction). Silk production is common among hexapod larvae. The most prolific silk producers, the Embiidina and Lepidoptera, have evolved dedicated silk glands. (Reprinted with permission of *Annual Review of Entomology*, Volume 42 © 1997, by Annual Reviews www.AnnualReviews.org.)

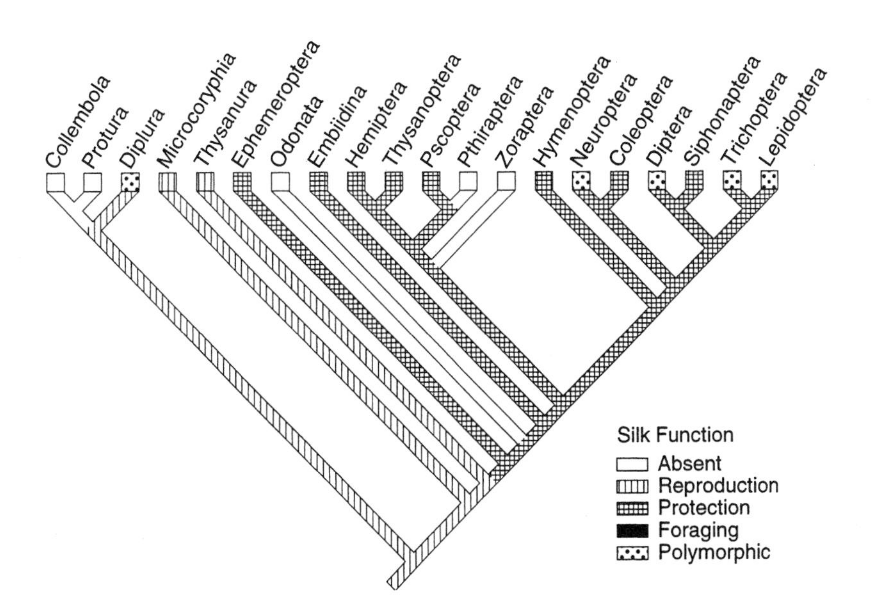

Figure 1.11 Function of silks produced by hexapods (one reconstruction). Silks were first used for reproductive purposes and later for protection. Only the Trichoptera (which make aquatic capture nets) and the Lepidoptera (which manipulate plant resources) have evolved to use silks in foraging. (Reprinted with permission of *Annual Review of Entomology*, Volume 42. © 1997 by Annual Reviews www.AnnualReviews.org.)

the insect orders analyzed included: (1) taxa that produce silks and fibrous proteins characterized by more than one macromolecular configuration; (2) taxa that produce silks and fibrous proteins as larvae and adults but in different glands; and (3) taxa that use silks and fibrous proteins for more than one purpose. When terminal taxa were polymorphic for a given character, all character states were coded (Nixon and Davis 1991). Despite polymorphism in the molecular configuration of silks and fibrous proteins produced by the Hymenoptera and Diptera and polymorphism in the types of silk-producing larval glands among the Hymenoptera, Neuroptera, and Coleoptera, in the majority of cases ancestral character tracing yielded only one most parsimonious tree.

The ability to secrete fibrous proteins is a primitive character of the hexapods and first evolved in the Diplura

Insects in almost all orders secrete some type of fibrous protein (figure 1.6), and most insect-secreted proteins are secondary products of organs that evolved for some other primary purpose (figure 1.8; Palmer 1990). The primitive hexapods, Diplura and Thysanura, secrete fibrous proteins from anal and colleterial glands (or accessory genital glands). The paurometabolous insects produce fibrous proteins—some identified as silks—from portions of their Malpighian tubules, colleterial glands, or peritrophic membranes. The fibrous proteins produced by the adult holometabolous insects (Lepidoptera, Diptera, Siphonaptera, Hymenoptera) are variable in structure, and all except one (Empidiidae: Diptera) are produced in colleterial glands. Larval holometabolous insects produce labial gland silks and fibrous proteins. These have been studied most extensively, probably because they are produced in large quantities and easy to collect.

For most insect groups, silks and fibrous proteins evolved as secondary products of insect reproductive and excretory systems and were first used for reproductive purposes (Richards and Davies 1977). For example, males in the primitive Apterygote order Thysanura (the bristletails) produce secretions that are drawn into threads used to restrict females during mating (Schaller 1971). Female mantids (Dictyoptera: Mantoidea), *Hydrophilus piceus* (Coleoptera: Hydrophilidae; Rudall and Kenchington 1971; Stehr 1987) and adult *Chrysopa flava* (Neuroptera: Chrysopidae) produce colleterial gland proteins from which they construct an egg sac stalk (Lucas and Rudall 1968; Rudall and Kenchington 1971; Sehnal and Akai 1990). With the exception of those produced by male Thysanura (Smith and Watson 1991), none of the colleterial gland proteins is delivered through a specialized spinning organ.

At least eight taxa of adult insects produce fibrous proteins in colleterial glands, indicating an early link to a mature reproductive physiology. Colleterial gland silk production has been lost and regained twice (figure 1.9). Among larvae, silk production in Malpighian tubules evolved early in the Ephemeroptera, then was lost and later reappeared sporadically across the insect orders (Neuroptera, Coleoptera, Hymenoptera). Labial silk glands evolved only twice, once in the Psocoptera and later in the derived, holometabolous insect larvae (figure 1.10).

Only the Embiidina and the Diplura have the ability to produce silks throughout their lifespan. The Diplura produce silks in their anal glands, and the Embiidina draw silks from specialized dermal glands. All of the derived holometabolous insect larvae produce silk proteins that they use for protection. This could be correlated with the fact that holometabolous larvae are soft bodied and, perhaps without silk, would be more vulnerable to predation.

With the exception of the Psocoptera, the only insects that produce labial gland silks are those whose development is holometabolous (figure 1.10). The Psocoptera have silk glands similar to those of caterpillars (Richards and Davies 1977). The Psocidae use silks to cover their eggs. Philotarsidae spin webs singly or gregariously in small groups, and Archipsocidae spin extensive webs covering trunks and branches of large trees (Stehr 1987; Smithers 1991). The labial gland silks and fibrous proteins have evolved in a few hemimetabolous Paraneoptera. Representatives of all endopterygote insect orders (Hymenoptera, Diptera, Siphonaptera, Trichoptera, and Lepidoptera), however, produce labial gland products when they are larvae. The tarsal glands of the Embiidina represent a unique evolutionary event in the history of the hexapods. The Embiidina are the only hemimetabolous insects that produce large volumes of silk, and they do so throughout their lives.

In summary, insect fibrous proteins were first produced in organs—anal and colleterial glands, Malpighian tubules, and peritrophic membranes—that had evolved for some other primary purpose. These fibrous proteins are simply secreted or deposited but not spun as defined above, and are often used with additional materials. The labial gland silks (the primary function of which is protection), produced by larval holometabolous insects, are spun. These more derived glands, dedicated to silk, produce proteins having different configurations. It is interesting to note that the Embiidina—although an order of phylogenetically primitive insects—produce spun silk proteins in specialized tarsal glands that seem to have evolved solely for this purpose. I will now compare and contrast the configurations of the more ancient and more derived fibrous proteins (figure 1.12).

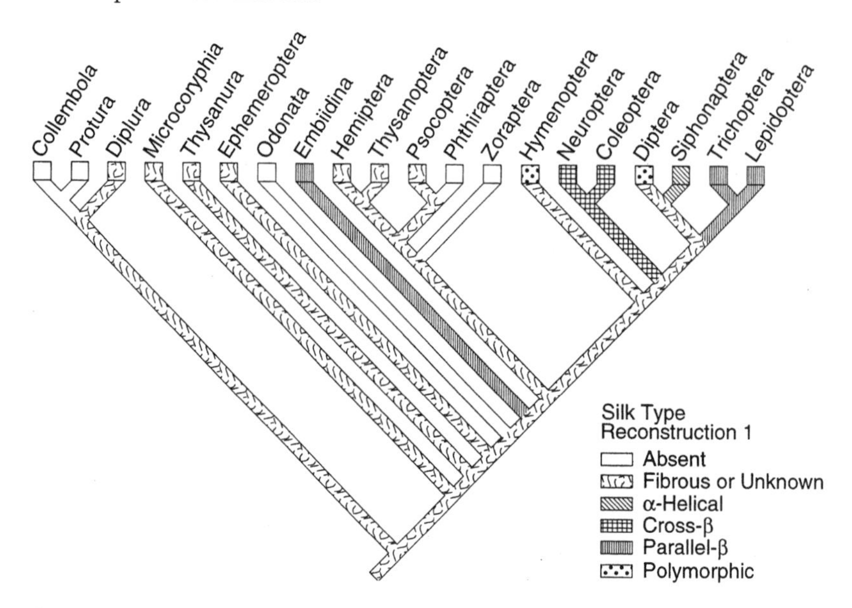

Figure 1.12 Structural configuration of silks produced by hexapods. In most cases, the structural configuration of silks produced by hexapods has not been determined. The Hymenoptera produce the greatest diversity of silks. Parallel-β silks only evolve in hexapods that are able to produce both their amino acid sequence but also a means of shearing the protein upon secretion. (Reprinted with permission of *Annual Review of Entomology*, Volume 42. © 1997 by Annual Reviews www.AnnualReviews.org.)

The ability to secrete silk fibroins correlates with the evolution of spinning behavior

Silks are spun when liquid proteins flow from a reservoir in which they are stored through a narrow constriction or nozzle (Kerkam et al. 1991). A pressure gradient between the reservoir and nozzle, plus tension applied to the silk as it is drawn or "spun," orients and shears the molecules, causing them to polymerize (Kerkam et al. 1991) into a fiber (Magoshi et al. 1994). Shear rate (= force $\cdot$ area^{-1} $\cdot$ time^{-1}) is a function of the speed at which the protein is pulled, but is indirectly affected by the morphology of the silk gland (for example, globular, tubular, sinuous) and the morphology of the spigot (length, cross-sectional area, and morphology of the aperture through which the protein is expelled). The animal's ability to control the shear rate influences the regularity of the thread surface, the diameter of the fiber, and the phase composition of the protein (Magoshi et al. 1994). Even though some insects, such as Diptera and Coleoptera, are able to produce β-protein structures, insects in neither of these groups have evolved behaviors that would shear the protein to result in the orientation and cross-linking of molecules into β-pleated sheet configurations.

Insects in only four different orders, the Lepidoptera, Trichoptera, Hymenoptera, and Embiidina, produce silk in the highly ordered, parallel-β configuration, and in each of these groups, a different method of spinning has evolved. Among the most derived, silk-spinning insects are the Lepidoptera larvae. They are the only insects capable of spinning silk proteins free from a substrate. They do so by a vigorous, repeated, figure-eight motion of the silkworm's head while the larvae's prolegs draw the viscous protein from the spinneret, stretching and shearing it (Magoshi et al. 1994). Although the Lepidoptera are able to regulate the release of the silks and shape of the emerging threads, they are unable to stop the flow of silk by any means other than biting or breaking the silk strand (Fitzgerald 1995). Larval weaver ants (*Oecophylla:* Hymenoptera) have not evolved the ability to spin silks independently. Mature ants hold silk-secreting larvae in their mandibles and shuttle them back and forth across leaf surfaces (Holldobler and Wilson 1990). By manipulating an individual larva that is releasing liquid silk, tension is applied to the protein streaming from larval spinnerets and the molecules are oriented into parallel-β fibroins. The resulting silk threads pull and bind nest leaves together. The third group of insects able to produce parallel-β silks is the Embiidina. Specialized dermal cells located on the insect's foretarsi are armed with setae-like ejectors. Silk proteins are expelled when the setae are swept across a stone, grass, or bark surface (Ross 1991) and the subsequent mechanical shear causes the silk protein to organize into the β-sheet configuration. Character reconstruction shows that with the exception of the silks produced by the Embiidina, all parallel-β fibroins are produced in dedicated, labial glands. Nevertheless, the labial glands also produce other types of fibrous proteins in α-helical and cross-β configurations.

The data indicate that both spiders and insects have the genetic systems and processing systems needed to produce silk proteins. They also have evolved a diversity of spinning behaviors that allow them to draw the silk from the gland under tension. Except for the Lepidoptera, all of the insects that produce silks make use of a substrate to spin. One end of the thread is fixed to a surface and then silk is drawn from the insect when it moves its body. Spiders use substrate spinning techniques, but they have also evolved "substrate-free" spinning, like the silk moths. In this case, the spider pulls silk from its spinneret using special claws at the end of each leg. Spinning speed and fiber manipulation affect fiber formation, suggesting that the evolution of the diversity of spider silks is correlated with their ability to manipulate the threads. Spiders are able to produce more types of silk with different functional properties than any other organism.

The diversity and retention of silk-secreting glands correlates with the phylogenetic status of web-spinning spiders

While the production of mixed protein threads has been found in only the most derived of the silk-producing insects, the Lepidoptera, even the simplest of the spider silk-producing systems make threads that contain multiple types of protein. (It is possible that further study may reveal mixed protein threads in other insects. However, the complexity of even the simplest spider systems is remarkable.) Investigation of the molecular conformation of silks produced by spiders has been limited to proteins drawn from the cylindrical (origin of egg-sac silks), ampullate (origin of dragline silks), flagelliform (origin of prey-capture threads spun by araneoids), and aciniform glands (prey wrapping and decoration silk) of only a few spider species (Lucas et al. 1960; Riekel et al. 1999a,b,c; C.L. Craig and C. Riekel, unpublished; see chapter 3). All of these proteins differ in the structure of their crystalline and probably noncrystalline domains.

The number of different silks produced by spiders and the physiology of their silk-processing systems seem to correlate with the phylogenetic status of spiders. The most complex, spider, silk-producing systems include some glands that are morphologically similar to the simple glands of ancestral spiders, in addition to glands that are both morphologically and physiologically complex (figure 1.13). Multiple glands allow spiders to produce more than one type of silk simultaneously. Mixtures of threads that include

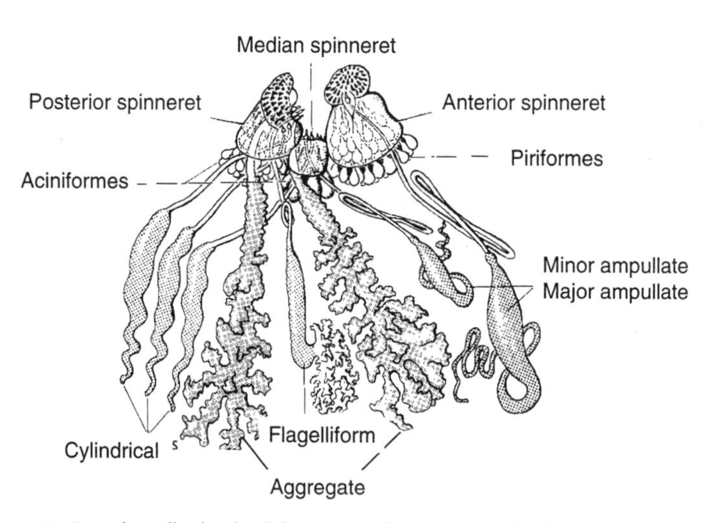

Figure 1.13 Complex silk glands of the araneoid spider *Nephila*. The araneoids have four glands from which they draw silk fibroins (major ampullate, minor ampullate, cylindrical, flagelliform) and two glands in which they produce protein glues, piriform and aciniform (not shown). (Modified from Peters 1955.)

fibers from ancestral and derived types of gland are not uncommon (C. Riekel and C.L. Craig, unpublished).

The most prolific silk-producing glands of the spiders are located in the opisthosoma (posterior portion of the body behind the cephalothorax or prosoma) and are tubular or acinous (grape-like) in form. The opisthosomal silk glands include the piriform and aciniform glands as well as the major and minor ampullate, aggregate, flagelliform, and cylindrical glands. All of these glands are dedicated to the production of silk and fibrous proteins and are thought to have evolved *de novo*; that is, independently of other organ systems (Palmer 1990) except, possibly, other silk glands. The evolution of glands for the production of silks and fibrous proteins in spiders is in marked contrast to insect silk glands, all of which result from the modification of some other organ. The presence of silk glands in all spiders throughout their lives dramatizes the primary importance of these proteins in their evolution.

The amino acid residues of parallel-β silk proteins drawn from the cylindrical, ampullate, and flagelliform glands are primarily glycine, serine, and alanine, which are all characterized by short, side-chain functional groups (Anderson 1970). In contrast, the piriform glands (common to all araneomorph or true spiders) and aggregate glands (found only in araneoid spiders) produce protein secretions characterized by a high content of basic amino acids, particularly lysine, and a high affinity for water (Anderson 1970). These silk polymers are unlikely to exhibit the α-helical configuration because of their high content of proline, and unlikely to assume a β-conformation because of their low content of amino acids with short side-chains (Szent-Gyorgyi and Cohen 1957). Therefore, they probably occur in a coiled state both before and after secretion, and it is therefore reasonable to infer that they are not pulled from the tip of the spinneret but are deposited. The piriform gland proteins, used to bind materials or cement threads to a substrate, dry quickly. Aggregate gland protein glues probably do not dry (Anderson 1970) due to their high concentration of hygroscopic, low molecular-weight compounds. These proteins are secreted in conjunction with flagelliform silk to make the catching thread sticky and enable silks to retain prey.

The fact that spiders do not abandon the "old" or less complex type of silk glands as new glands evolve indicates that their multigland system evolved through selection for increased silk investment and not just selection for new or alternative silk function. A correlate of this fact is that the web spinners expend much more metabolic energy on silk production than related spiders that do not use silks for foraging. As an alternative to the functional hypothesis (or perhaps in addition to it), divergence in gland types and processing systems may result from intra-gland competition for limited amino acids (chapter 7). New processing physiologies are also needed to make use of alternative amino acid complexes. It may be significant that the fibrous proteins produced by ancestral spiders whose respiratory systems are limited contain a greater proportion of amino acids gathered from the environment, while the amino acids in silks and fibrous proteins produced by more derived spiders can be synthesized.

The spinnerets through which these fibrous proteins are spun are located on the fourth and fifth opisthosomal segment of the true spiders. Acinous, epiandrous silk glands, common to most males, are located on the anterior rim of the abdominal, epigastric furrow and open through simple spools. The histochemical characteristics of the epiandral glands are the same for all spiders that have been studied to date (Kovoor 1987), but the characteristics of all other glands are variable among spider taxa.

There is one exception to the idea that spider glands arose *de novo:* the modified cheliceral glands of spiders in the family Scytodidae, which produce fibrous proteins of undetermined structure. These glands are homologous to insect labial glands (Kovoor 1987) thought to be derived from labial glands. It may be significant that in the Scytodidae, abdominal spigots (and presumed glands) are highly reduced to include primarily piriform and aciniform spigots (Platnick et al. 1991). Prosomal glands (glands in the head and thorax or prosoma) have also evolved independently in the other arachnids Acariformes as well as some Pseudoscorpiones (Kovoor 1987; Hunt 1970; Adis and Mahnert 1985).

All spiders secrete simple, fibrous proteins; some spiders produce silks characterized as composite materials that contain multiple functional domains

Individual spider species have been labeled either generalists or specialists on the basis of their silk-producing systems. Generalist spiders are those that produce one to three kinds of fibers but use them indifferently for the construction of burrows, egg sacs, and sperm webs (Kovoor 1987). Specialist spiders produce four to nine kinds of silks and fibrous proteins (distinguished by their glandular origin), each of which they use specifically to construct burrows, egg sacs, sperm webs, or feeding nets. The silks of specialist spiders are characterized by mixed fiber compositions and may include more than one type of protein as well as multiple protein domains. At present, it cannot be stated with certainty that spiders do not substitute fibroin silks for one another under some conditions.

The Mesothelae (Haupt and Kovoor 1993), generalist silk producers, are the most primitive living spiders. Anatomical and histological studies show that they produce three different fibrous proteins that they use indiscriminately to line their burrows or cover their eggs (Haupt and Kovoor 1993). The Mesothelae gland system would seem to indicate the spiders' need for a particular volume of fibrous proteins but not for any particular type of fibrous protein or silk (but see chapter 3).

The Opisthothelae, including both the Mygalomorphae (that is, tarantulas and trap-door spiders) and the Araneomorphae (or "modern" spiders), produce fibrous proteins and silks composed of more than one type of protein. The simplest spider silk producing system, that of the mygalomorph *Antrodiatus unicolor* (Palmer et al. 1982), is characterized by only one type of gland, an acinous gland, and produces two types of proteins secreted through spigots on two pairs of spinnerets (Palmer et al. 1982). The distal cells of the glands produce a basic protein rich in sulfhydryl groups that may function in the cross-binding of the protein's crystallites. These proteins have a double composition consisting of a core and outer coating (Kovoor 1987).

The structural organization of spider silk is correlated with the evolution of a muscular and innervated spinneret

Silks are delivered in two ways, through cuticular spigots (present in all insects and some spiders) or through a flexible spinneret and spigot (present in no insects but in all spiders; figure 1.14). Some types of spider spinnerets—for example those that serve the specialized labial, cribellar, and epiandral glands—are derived from hollow setae and thus are similar to the fixed cuticular extensions or epidermal modifications that are characteristic of all insect spinnerets as well as those of the Acarina and Pseudoscorpiones (cited in Kovoor 1987; Schultz 1987). However, the key feature that distinguishes spiders from all other silk-secreting arthropods is the muscular, innervated spinnerets that serve the opisthosomal silk glands that the mygalomorph and araneomorph spiders hold in common. The sensory capability of these, their ability to move independently of the spider's body, and the spider's ability to adjust the size of the spigot opening allow great control over the rate at which the proteins are released and the properties of the threads that are spun. As a result, spiders have great flexibility with respect to the types and character of the threads they spin. Insects are constrained to produce a much more limited kind of thread because their spinnerets are neither muscular nor innervated.

As has been previously discussed, the production of highly ordered, parallel-β silk proteins requires the confluence of three events: the evolution of the protein, the evolution of a processing system, and the evolution of a spinning behavior. Spinning behavior is defined here as a means of pulling the liquid protein from the

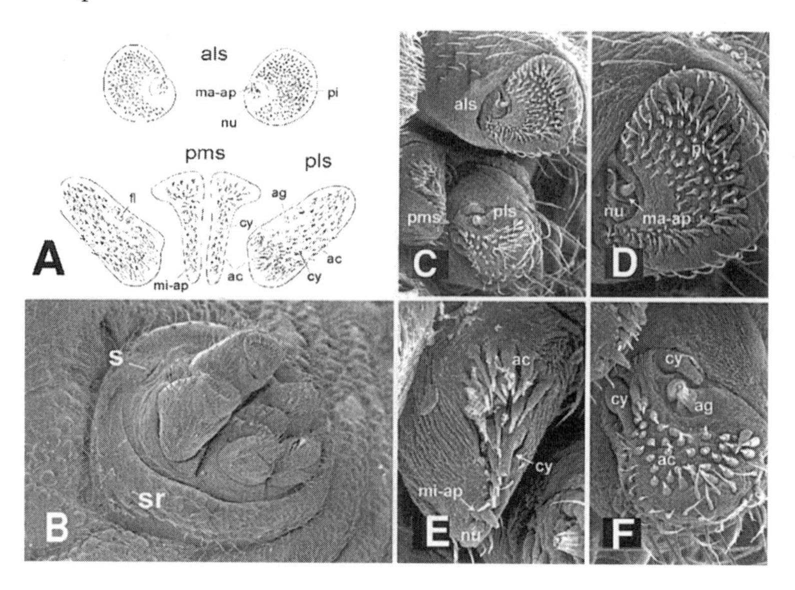

Figure 1.14 The unique spinnerets of spiders are muscular and innervated and are expressed from multiple spigots of diverse morphology. (A) Generalized diagram of spinnerets and spigots (ac = aciniform gland spigots; ag = aggregate gland spigots; als = anterior lateral spinnerets; cy = cylindrical gland spigots; fl = flagelliform gland spigots; ma-ap = major ampullate gland spigot; mi-sp = minor ampullate gland spigots; nu = nubbin; pi = piriform gland spigots; pls = posterior lateral spinnerets; pms = posterior median spinnerets. (From Coddington 1989). (B) The complex types of silks produced by the araneoid orb spinner, *Gasteracantha geminata,* are deposited or drawn from three pairs of spinnerets: the anterior, posterior, and median (s = spiracle; sr = sclerotized ring). (C–F) The multiple spigots from which fibrous proteins and silks are expressed. Some spinnerets have multiple spinning fields of morphologically distinct spigots that are fed by different glands. (Reprinted from Scharff and Coddington 1997, *Zoological Journal of the Linnean Society* with permission from Blackwell Press.)

gland, causing the β-structures to align and cross-polymerize. Of the insects, the Lepidoptera have evolved the most complex uses of silks—for communication (Fitzgerald and Peterson 1988), manipulating their environment (Berenbaum et al. 1993), and protection (Ruehlmann et al. 1988). Once the protein evolved, further functional modifications may not have been important due to the similar mechanical requirements of lepidopteran silks (Palmer 1990). In contrast, the derived, aerial web-spinners vary over an order of magnitude in body size and produce nets that also vary over an order of magnitude in size and kinetic properties (Craig 1987a; Craig 1987b). All of the aerial web spinners have evolved a rich repertoire of spinning behavior that determines the rate that the proteins are sheared and contributes to their material properties. In fact, consistent variations in spinning behavior across spider systematic groups have resulted in their use for systematic analyses (Coddington 1989;

Coddington 1990a). Spinning behavior seems a likely target for evolution by natural selection.

Silk production in insects and spiders is hypothesized to have evolved via two pathways: a systemic gland pathway and a surficial gland pathway

In 1932, Bristowe proposed that spider spinnerets evolved from modified leg segments and the spigots, through which silks are delivered, are derived from simple hairs (Bristowe 1932). More recent studies have suggested that silk-producing glands of spiders are derived from epidermal invaginations (Kovoor 1987; Palmer 1990). This view has been supported by the proposal that the protein-secreting, epiandral glands of most male spiders are derived from dermal glands (Kovoor 1987). Several investigators have proposed that the epiandral glands are homologous with the "true" silk glands of spiders, the opisthosomal, or abdominal, glands (reviewed in Shultz 1987). This hypothesis has been further extended to suggest that the silk-producing systems of spiders are most similar to the dermal silk-producing systems of the Embiidina (Palmer 1990). A third hypothesis proposes that the glands are epidermal invaginations but suggests that the spigots originated as sensory hairs, not simple hairs. This proposal is based on the fact that the silk spigots of mygalomorph spiders are morphologically similar to chemoreceptors and are characterized by well-developed slit sensilla at their base (Palmer 1990).

All silks derive from an ectodermal cell lineage, but the pathway to fibroin silk glands and glands that secrete protein glues may be different. A fourth hypothesis was proposed relatively recently, that silk glands evolved via two different pathways in insects and possibly spiders as well (Craig 1997). These are a surficial cell pathway leading to the evolution of fibrous proteins, and a systemic pathway leading to the evolution of fibroin silks. An example of evolution along the surficial pathway would be the silk glands that develop from epidermal invaginations and dermal cell modifications, such as the epiandral glands of spiders and the silk glands of the Embiidina. A systemic pathway would be exemplified by the modified salivary gland systems of Lepidoptera and the major and minor ampullate glands, cylindrical or flagelliform glands of spiders. Insects have either systemic or surficial silk-producing pathways; some mygalomorph and all araneomorph spiders retain both.

The morphological similarities of (1) the simple undifferentiated glands of the liphistiomorph and mygalomorph spiders; (2) the differentiated piriform, acinous, cribellar, and epiandral silk glands of the araneomorph spiders; and (3) the dermal glands of the Embiidina

(Empiidae: Diptera) and some Hemiptera glands suggest that they are derived from a surficial silk-producing system. The proteins produced in these glands appear fibrous, but not all exhibit the highly ordered, parallel-β configuration. Silk production via a surficial pathway is possible among arthropods of all development types, and dermal silk-producing glands have evolved in both adults and larvae (figure 1.15).

All of the silk glands that seem to have evolved via a systemic pathway (modified organ in insects or *de novo* in spiders) produce proteins that contain a parallel-β configuration in some component. These include silks produced in the major and minor ampullate, cylindrical, and flagelliform glands of spiders and the labial glands of the holometabolous insect larvae. All of these glands are characterized by an elongated, and in some cases sinuous, morphology that is likely to be important in protein processing and storage. It may be significant that these glands are characteristic of more derived silk producers. Additional fibrous proteins are produced via other, secondary, systemic pathways. Fibrous proteins produced by these pathways are characterized by α-helical or cross-β configurations. In these cases, silks are produced in a region of a gland that serves some other primary function. Secondary systemic systems include the Malpighian tubules and the colleterial and anal glands found in more ancestral insects (Richards and Davies 1977).

PATTERNS OF SILK-PRODUCING SYSTEMS & PROPOSED EVOLUTIONARY PATHWAY

Ectodermal Cell Lineage

	Possible Surficial Glands		**Possible Systemic Glands**	
	Paurometabolous	Holometabolous	Paurometabolous	Holometabolous
Insects	Adults & Larvae **tarsal glands** (Embiidina)	Adults **tarsal glands** (Empiidae: Diptera)	Adults & Larvae *salivary glands* (Orthoptera) *colleterial glands* (Mantoidea)	Adults *colleterial glands* (Hymenoptera) Larvae ***salivary glands*** (Lepidoptera) *peritrophic membrane* Adults & Larvae *Malpighian tubules* (Coleoptera)
Spiders	Adults & Larvae **piriform, acinous,** (Araneomorphae) **cribellar glands** (Deinopoidea)	None	Adults & Larvae **tubulliform,** **ampullate,** (Araneomorphae) **pseudoflagellar,** (Uloboridae) **flagellar glands** (Araneoidea)	None

Glands indicated in bold are dedicated to the production of silk;
glands indicated in italics produce silk as a secondary function.

Figure 1.15 Possible evolutionary pathways for the evolution of silk production in insects and spiders. (Modified from Craig 1997.)

Embryological data show that spinnerets are homologous with other arachnid prosomal appendages, including chelicerae, pedipalps, and walking legs (Dawydoff 1949). Here, spinnerets are proposed to have arisen from segmented appendages later modified for silk delivery (Schultz 1987). The lateral spinnerets of the spider *Liphistius* spp., however, are not similar to the segmented appendages of arthropods but instead are more similar to the lobopod appendages of the Onychophora (Palmer 1990).

Summary

Secreted, fibrous proteins and silks are produced only by arthropods. Specifically, they are produced by at least one representative in all of the arthropod orders except the Crustacea. All silk-producing cells are from an ectodermal cell lineage. Current systematic analysis suggests that silks evolved independently in the Arthropoda, while molecular genetic analysis suggests that silk expression is the product of a homeotic gene. In this chapter, I have compared the molecular organization and glandular organization of silks and fibrous proteins and suggested that the protein-secreting glands evolved via three pathways. One pathway is a "surficial" gland pathway, meaning that fibrous proteins and silks are produced in modified ectodermal cells and that silk production is independent of type of development. The second pathway is a systemic pathway. Fibroins and fibrous proteins are produced in organs modified from some other primary purpose, as in the case of the salivary gland silks produced by the Lepidoptera larvae. Finally, some silk glands evolved *de novo*, as with the ampullate or from other silks glands, such as the pseudoflagelliform glands in spiders.

Comparison of the molecular properties of fibrous proteins and silks shows that their functional properties and structural configurations are diverse. Both structure and function are related to the amino acid compositions of the silks. However, silk structural organization is also related to the physical conditions, such as pH and temperature, under which the silks are produced. In some cases, the structural configurations of fibrous proteins are the result of silk spinning behavior, or whether the silks are deposited or pulled and sheared at secretion. Despite this flexibility in structure, the organization of the complex silk proteins, the derived fibroins, is also dependent on the presence of four, specific types of amino acids: glycine, alanine, serine, and proline. Less complex silks or fibrous proteins are more variable in amino acid composition.

The specific types of amino acid that make up silks and the organization of the protein molecule are ultimately determined by the molecular genetic architecture of the silk genes and the regulation of gene expression. The observed diversity of silks and complex nature of the composite silk fibroins suggest that their evolution may also be complex. For example, does the structural conformity of silk fibroins and proteins derive from the same molecular genetic architecture or can different molecular genetic architectures yield similar silk proteins? How easily do different types of silks evolve? What are the structural constraints on the evolution of new silk functional properties and how are the specific confirmations of proteins maintained during protein transcription? The interaction between fibrous protein and silk molecular organization and their molecular genetic architecture is discussed in the next chapter.

2

The Comparative Architecture of Silks, Fibrous Proteins, and Their Encoding Genes in Insects and Spiders*

with C. Riekel

Evolvability is the capacity of the genome to produce new genotypes (Kirschner and Gerhard 1998). Whether gene changes and new interactions result in adaptive change depends on how tightly the genotype is mapped onto an organismal phenotype (Wagner and Altenberg 1996). One of the most direct genotype-phenotype maps is that charting the relationship between the molecular genetic organization of the silk gene and the molecular organization and function of silk proteins. As in venoms, plant and defense compounds, and proteinaceous flower pigments, the gene product is used externally in a way that is easily measured.

The two most important periods of speciation in the Araneae—the divergence of the Araneomorphae from the Mygalomorphae in the Middle Triassic (Seldon and Gall 1992) and the divergence of the derived Araneoidea from their ancestral taxa the Deinopoidea during the early Cretaceous (Seldon 1989)—correlate with the evolution of two types of silk-producing glands: the major ampullate gland (*MA*) and the flagelliform gland (*Flag*). Despite the importance of these events, little is known about the selective and molecular mechanisms that resulted in the evolution of the *MA* and *Flag* silk proteins. In this review we will compare the molecular genetic organization of *MA* and *Flag* genes and proteins produced by spiders to two types of

*Reprinted from Comparative Biochemistry and Physiology Part B, *133* 493–507 (2002) with permission of Elsevier Press.

fibroin silks produced by Lepidoptera larvae: the heavy chain fibroin (*Fhc*) spun by the larval moth *Bombyx mori* (*Bm-Fhc*; Bombycidae) and that spun by its sister taxa *Antherea pernyi* (*Ap-Fhc*; Saturniidae). In addition, we will discuss the molecular genetic organization of the fibrous glue sericin (*Ser*; produced by *Bombyx mori*) and the Balbiani Ring (*BR*) gene proteins produced by the larval fly *Chironomus tentans* (Diptera) as possible models for the fibrous protein glues produced by spiders.

The fibroin genes and proteins differ at the level of their molecular genetic architecture, in amino acid sequence, and in structural configuration of the protein. In addition, all of the silk genes are inherently prone to recombination errors due to their highly repetitive nature and codon biases, and display considerable allelic variation that is attributed to a high frequency of unequal cross-over (Gage and Manning 1980; Mita et al. 1994; Hayashi and Lewis 2000; Sezutsu and Yukuhiro 2000). Despite their architectural differences, sequence differences and instability, there are emergent properties of the organization of silk genes and proteins that suggest that a dynamic evolutionary conflict between genetic processes and natural selection may have played a role in silk evolution (Hayashi and Lewis 2000; Gatsey et al. 2001).

The known silk fibroins and fibrous glues are encoded by members of the same gene family

The early divergence of the Araneomorphae (150 mya) and the Araneoidea (65 mya) from their sister taxa correlate with the evolution of *MA* and *Flag* silks. Despite their ancient roots, the repetitive nature of spider silk proteins is largely attributed to the variable proportions of the four amino acid motifs A_n, GA, GGX, and $GPG(X)_n$. While the sequence $GPG(X)_n$ is unique to the *MA2* and *Flag* silks produced by the araneoids, even the silks and fibrous glues of Lepidoptera larvae contain A_n, GA and GGX components (table 2.1). Therefore, despite the fact the silks spun by insects and spiders evolved independently from one another, they display some remarkable similarities that could suggest convergence and stabilizing selection have been important factors in their evolution (Gatsey et al. 2001).

Within the Lepidoptera and Araneidae, all silk proteins and fibrous proteins are proposed to be members of a common gene family (Mita et al. 1994; Guerette et al. 1996; Hayashi and Lewis 1998). The silk proteins these genes encode vary in structural organization, but in general are made up of different combinations of β-pleated sheets tightly packed into crystalline arrays, loosely associated β-sheets, α-helices, β-spirals, and spacer regions of anom-

Table 2.1 Cross-taxa comparison of repetitive sequences of silks.

	Primary repetitive motifs*	Source	Ref
Bombyx mori (Bm-Fhc)	$(GGNGCN)_n$GGTTCW n = 0 – 6, W = A or T, N = A, G, T, or C $GGNGYN)_n$GGNTAY n = 0 ~ 8, Y = T, C	Genomic DNA, complete	(Zhou, Confalonieri et al. 2000)
Antheraea pernyi (Ap-Fhc)	Constant domains $GGYGSDS(A)_{12}$GSGAGG $GGYGSGSS(A)_{13}$SGAGG Variable domains AGGGYGWGGD $A(GGY)_n$ RGD RRAGHDRAAGS	Genomic DNA, complete	(Sezutsu and Yukuhiro 2000)
Nephila clavipes(Ma-1)	Possible domain organizations $(A)_N$GGA(Repetitive with variation)GAG(Variable)	cDNA, partial	(Xu and Lewis 1990)
Nephila clavipes (Ma-2)	Possible domain organizations $(A)_N$(variable)PGGYGPGQQGGPGGY(Variable)	cDNA, partial	(Hinman and Lewis 1992)
Nephila clavipes (Flag)	$GPGG(X)_n(GGX)_n$(Variable and nonrepetitive)$(GGX)_n(GPGGX)_n$	Genomic, partial cDNA, partial	(Hayashi and Lewis 2000)
B. mori (SerI)	LSEDSSEVDIDLGGNLGWWWNSDNKAQRAAGGATKSEASSSTQ $(SRTSGGTSTYGYSSSHRGGSVSSTGSSSNTDSSTKNAG)_n$	Genomic, cDNA	(Garel, Deleage et al. 1997)
Chirnomous tentans(Br1)	AAATCTGGACCAAGATCAAAGC AAACCTGAAAAGACCAAGCAAATCAGGACCT AAACCAAGTAAGGGATCTAAACCTAGACCAGAG	Genomic, partial cDNA, partial	(Paulson, Hoog et al. 1992)

* Primary repeat motifs are abstracted from full sequences. See reference for complete sequences and polymorphic sites.

alous amino acid sequence (Xu and Lewis 1990; Beckwitt and Arcidiacno 1994; Guerette et al. 1996; Simmons et al. 1996; Hayashi and Lewis 1998).

The *Fhc* silk spun by *B. mori* is the product of three genes. The first identified silk gene encodes for a very large, insoluble protein designated as the heavy-chain fibroin (*Fhc*, 350 kDa). Two additional genes, the L-chain (25 kDa designated as *Fl*) and the P25 protein, encode two subunits of the *B. mori* fiber that are linked to the fibroin by disulfide bonds. Together, the protein products of these three genes make up the water-insoluble silk core of the cocoon silk fiber (Grzelak 1995).

B. mori also produces sericin, a protein glue that ensures that the fibroin cocoon threads adhere to one another. Five sericin genes (*Ser1*, *Ser2*, *MSGS3*, *MSGS4* and *MSGS5*) have been cloned and partially sequenced (Grzelak 1995). The structure of the *Ser1* gene is best known. It encodes six mRNA transcripts that result from selective splicing of the primary mRNA transcript (Garel et al. 1977; Michaille et al. 1986; Couble et al. 1987; Hayashi and Lewis 1998). Selective splicing of mRNA transcripts may also be an important mechanism for producing alternative proteins with alternative properties in other silk proteins. All of the fibroin proteins and *Ser1* share homologous 5′ flanking sequences (Grzelak 1995).

The Diptera fly larva *C. tentans* secretes fibrous proteins that it uses to construct an underwater tube from which it feeds. Five genes make up the Balbiani ring (BR) gene proteins (*BR1*, *BR2.1*, *BR2.2*, *BR3*, *BR6*). Four of these encode an uninterrupted block of 100 repeat units that are tandemly arrayed and almost identical. The gene *BR3*, however, is characterized by a different structural organization. *BR3* encodes a 10.9 kb transcript that is spliced into 5.5 Kb mRNA. Fifty-eight introns separate coding units that vary in size between 17 and 678 bp. However, the positions of the introns relative to the repeat structure of the protein are the same.

Adult female araneoid spiders have seven different types of glands that yield four fibroin silks and three types of protein glue (Kovoor 1987; Craig 1997). The fibroin silk genes in spiders are expressed in the major ampullate gland (*MA*), the minor ampullate gland (*MiA*), the flagelliform gland (*Flag*), and the cylindrical gland (*Cy*). Protein glues are produced in the piriform gland (*Pir*), the aggregate gland (*Ag*), and the aciniform glands (*Ac*). Aciniform gland silk has some crystalline organization; the physical structure of the piriform proteins is unknown. Recent X-ray diffraction patterns of aggregate gland proteins suggest that they may also contain some crystalline β-sheet material (C.L. Craig and C. Riekel, unpublished).

Both the Lepidoptera and Diptera produce silks or fibrous proteins in modified salivary glands when they are larvae. Spiders produce silks in multiple glands and throughout their lives. The differences in number of silks, gland morphology, and life stages in which production occurs suggest that there have been different selective factors acting on Lepidoptera and spiders that affect silk organization and function. For example, in spiders, the greater diversity of proteins is the result of divergence among the silk genes as well as the evolution of specialized protein-processing systems (Vollrath and Knight 2001).

Most silk fibroins contain crystalline and noncrystalline regions

X-ray diffraction and NMR data show that the *MA* silks spun by *Nephila* and *Fhc* silks spun by *B. mori* and *A. pernyi* contain crystalline and noncrystalline components (figure 2.1). The proposed molecular organization of the *MA* genes is similar to that of the *Ap-Fhc* gene, but both are fundamentally different from that of *Bm-Fhc*. The crystalline domains of the *Bm-fhc* fibroins are largely composed of repeating glycine and alanine that are arrayed in two related patterns. *Ap-fhc*, *Ma1*, and *Ma2*, in contrast, contain strings of poly-alanine that result in their crystal-forming domains (table 2.1).

In all of these silks, crystalline regions are interspersed by domains of 34–40 amino acids that make up the noncrystalline regions of the protein. Different silks, however, have different proportions of noncrystalline and crystalline fractions. About 40–50% of *B. mori* silk is made up of protein crystals (Iisuka 1965). When hydrated, only about 15% of the total volume of MA silks is crystalline (Gosline et al. 1999) which can be related to the swelling of the noncrystalline fraction. X-ray diffraction data on *Nephila* silk suggest a size of <6 nm for the crystalline β-sheet domains (Grubb and Jelinski 1997; Riekel et al. 1999a). More complex models were developed based on analytical transmission electron microscopy. Thus, the existence of large β-sheet crystallites of 70–100 nm size was proposed for *Nephila* silk (Thiel et al. 1994). In related experiments, the notion of nonperiodic lattice crystals (NPL) was introduced and the possibility that the β-sheet structure could change locally from crystalline to aperiodic configuration (Thiel et al. 1997; Thiel and Viney 1997). X-ray diffraction data (Thiel et al. 1997) suggesting the existence of more complex structural features than the classical β-sheet structure, however, have not been confirmed by single fiber X-ray diffraction (Riekel et al. 1999a).

The noncrystalline fraction of silk is currently the focus of many investigations of silk structure. For example, *MA* silk contains at least two different types of proteins that differ in their amino acid

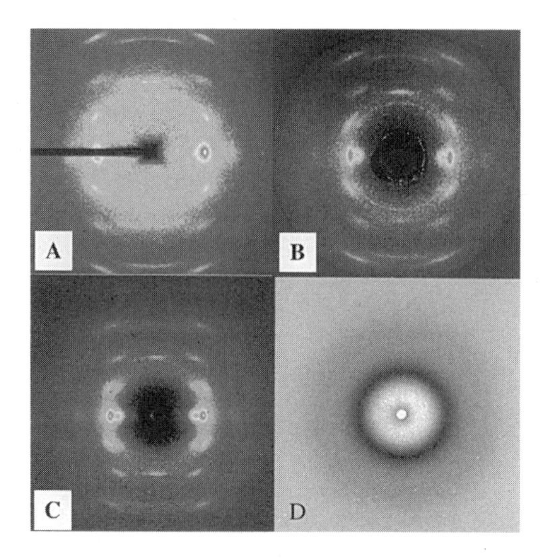

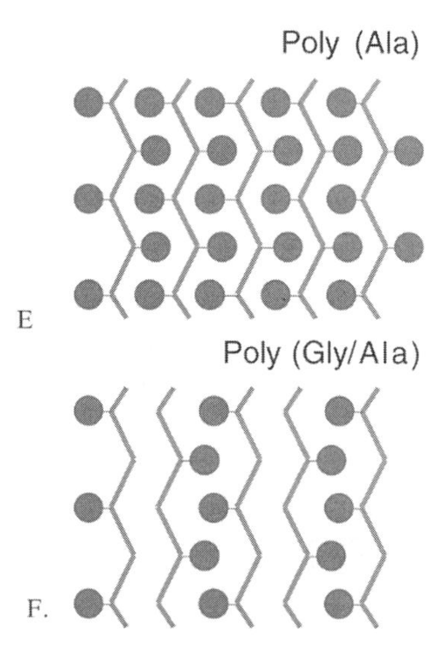

Figure 2.1 X-ray diffraction patterns of *Fhc* and *MA* silks spun by (A) *Bombyx mori,* (B) *Nephila clavipes,* (C) *Argiope argentata* show crystalline and noncrystalline regions; X-ray diffraction patterns of *Flag* silk spun by *Micrathena gracilis* show no crystalline regions (D). The fiber diffraction patterns of *N. clavipes* and *A. argentata MA* silk illustrate a poly-alanine crystal organization, while the fiber diffraction pattern from *B. mori* cocoon silk illustrates a poly(glycine-alanine) crystal organization. (E, F) Diagrams showing how poly-A and poly-GA amino acid organization affects crystal packing. The diffraction pattern for *M. gracilis Flag* silks shows only an amorphous halo.

content (figure 2.2). The cDNA designated as *Ma1* is made up of repeated motifs largely composed of glycine and alanine. The cDNA designated as *Ma2* is made up of repeated motifs largely composed of the amino acids glycine, alanine, and proline. Thus, although *Ma1* contains almost no proline, proline makes up about

A. cDNA for Major ampullate protein 1 *(Ma1)*. Possible genetic hot-spot in bold. Note relative paucity of amino acid proline.

Q**GAG**
AAAAAA**GGA**GQGGYGGLGGQGAGQGGYGGLGGQGAGQGAG
AAAAAAA**GGA**GQGGYGGLGSQGAGRGGQGAG
AAAAAA**GGA**GQGGYGGLGSQGAGRGGLGGQGAG
AAAAAAA**GGA**GQGGYGGLGNQGAGRGGQG
AAAAAA**GGA**GQGGYGGLGSQGAGRGGLGGQGAG
AAAAAA**GGA**GQGGYGGLGGQGAGQGGYGGLGSQGAGRGGLGGQGAG
AAAAAAA**GGA**GQGGLGGQGAGQGAGAS
AAAA**GGA**GQGGYGGLGSQGAGRGGEGAG
AAAAAA**GGA**GQGGYGGLGGQGAGQGGYGGLGSQGAGRGGLGGQGAG
AAAA**GGA**GQGGLGGQGAGQGAG
AAAAAA**GGA**GQGGYGGLGSQGAGRGGLGGQGAGAV
AAAAA**GGA**GQGGYGGLGSQGAGRGGQGAG
AAAAAA**GGA**GQRGYGGLGNQGAGRGGLGGQGAG
AAAAAAA**GGA**GQGGYGGLGNQGAGRGGQG
AAAAA**GGA**GQGGYGGLGSQGAGRGGQGAG
AAAAAAV**GAG**QEGIRGQGAGQGGYGGLGSQGSGRGGLGGQGAG
AAAAAA**GGA**GQGGLGGQGAGQGAG
AAAAAAGGVRQGGYGGLGSQGAGRGGQGAG
AAAAAA**GGA**GQGGYGGLGGQGVGRGGLGGQGAG
AAAA**GGA**GQGGYGGVGSGASAASAAASR
LSPQASSRVSSAVSNLVASGPTNSAALSSTISNVVSQIGASNPGLSGCDVLIQALLEVV
SALIQILGSSSIGQVNYGSAGQATQIVGQSVYQALG"

B. cDNA for Major ampullate protein 2 *(Ma2)*. Possible genetic hotspot (PGG) in bold.

PGGYGPGQQGPGGYGPGQQGPSGPGS
AAAAAAAAAAG**PGG**YGPGQQGPGGYGPGQQGPGRYGPGQQGPSGPGS
AAAAAAGSGQQG**PGG**YGPRQQGPGGYGQGQQGPSGPGS
AAAASAAASAESGQQG**PGG**YGPGQQGPGGYGPGQQGPGGYGPGQQGPSGPGS
AAAAAAAASGPGQQG**PGG**YGPGQQGPGGYGPGQQGPSGPGS
AAAAAAAAS**GP**GQQG**PGG**YGPGQQGPGGYGGQQGLSGPGS
AAAAAAAGPGQQG**PGG**YGPGQQGPSGPGS
AAAAAAAAAG**PGG**YGPGQQG**PGG**YGPGQQGPSGAGS
AAAAAAAGPGQQGLGGYGPGQQG**PGG**YGPGQQG**PGG**YGPGSAS
AAAAAAGPGQQG**PGG**YGPGQQGPSGPGSAS
AAAAAAAAG**PGG**YGPGQQG**PGG**YAPGQQGPSGPGSAS
AAAAAAAAG**PGG**YGPGQQG**PGG**YAPGQQGPSGPGS
AAAAAAAAAG**PGG**YGPAQQGPSGPGIAASAASAG**PGG**YGPAQQGPAGYGPGSAVAASAGAGSAGYGPGSQ
SAAASRLASPDSGARVASAVSNLVSSGPTSSAALSSVISNAVSQIGAS**NPG**LSGCDVLIQALLEIVSACV
ILSSSSIGQVNYGAASQFAQVVGQSVLSAFX

Figure 2.2 Amino acid sequence of cDNA for *Ma1* (A) and *Ma2* (B) silks produced by *Nephila clavipes*. The cDNA shows that proline is virtually absent from the *Ma1* cDNA sequence although it is abundant in the *Ma2 cDNA* sequence. Furthermore, both *MA* sequences contain DNA "hot-spots" (GAA, *Ma1*; PGG, *Ma2*). The arrangement of the *MA* PAB (crystalline) and NPAB (noncrystalline) domains and may be similar to the structure proposed for the silk fibroin *A. pernyi*.

15–17% of the repeat motif in *Ma2* (Gosline et al. 1999). Proline is thought to affect the organization of the noncrystalline regions of the proteins and specifically silk elasticity (Hayashi and Lewis 1998; Gosline et al. 1999).

Using NMR analyses, van Beek et al. (van Beek et al. 2000) proposed that the glycine-rich, noncrystalline domains of *Ma1*, a proline-free silk, are organized into aggregated, 3_1-helix or Type I β turns that reinforce the highly oriented, polyalanine polymer network. X-ray diffraction data suggest the presence of two types of noncrystalline material: (1) randomly oriented material; and (2) oriented material (Grubb and Jelinski 1997; Riekel et al. 1999a; Riekel et al. 1999b). The oriented noncrystalline fraction has been suggested to form a fibrillar system with the crystalline fraction (Yang et al. 1997; Riekel and Vollrath 2001). Nevertheless, the model of NPL crystallites (see above) is an interesting hypothesis but has not been verified by other groups.

The models of the two silks suggest that while proline enables the large-scale organization in *Ma2*, it disrupts the structural organization of the glycine-rich helices in *Ma1* (Gosline et al. 1999). Furthermore, the conserved, poly-alanine sequences that make up the crystalline regions of *Ma1* and *Ma2* in conjunction with the divergent nonpolyalanine sequences that make up the noncrystalline regions of the silks may suggest that any diversification in *MA* silk structure and function is the result of variation in the noncrystalline regions of *MA* proteins. The noncrystalline regions of *MA* silks are probably the primary site of evolutionary innovation (Craig et al. 2000).

Silks produced in the flagelliform glands of spiders do not contain polyalanine, nor do they contain domains of β-pleated sheets. In the few samples of Flag silk so far examined, X-ray diffraction has not shown a crystalline fraction (figure 2.1D; chapter 3).

All of the sequenced fibroin silks (*Fhc, MA,* and *Flag*) are made up of hierarchically organized, repetitive arrays of amino acids

The cocoon silks produced by the Lepidoptera moth larvae *B. mori* (Bombycidae) (*Bm-Fhc*) and *Antheraea pernyi* (*Ap-Fhc*), spider *MA* silk, and *Flag* silk are made up of repetitive amino acid sequences that are highly conserved both within and, in some cases, across taxa (figures 2.2–2.6; table 2.1). Cocoon silks produced by *B. mori* are made up of sequences of glycine and alanine $(GA)_n$ organized into tandem units of GAGAG(S) and GAGAGX (X varies between tyrosine and valine; Zhou et al. 2000). Even though they are sister taxa, cocoon silk sequences from *A. pernyi* (*Ap-fhc*) (Saturniidae) are fundamentally different from *B. mori*

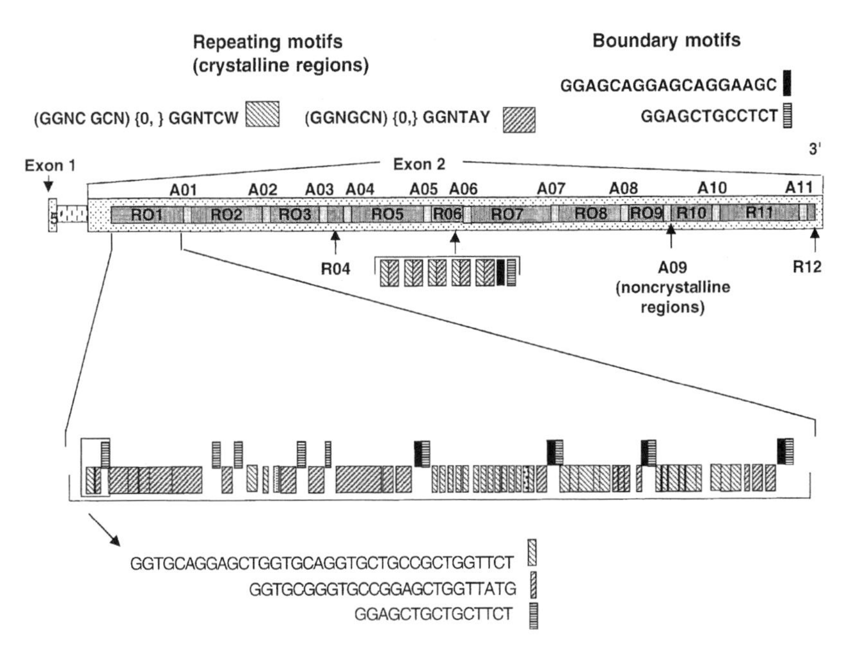

Figure 2.3 The repetitive sequences of *Fhc* silks spun by *B. mori* are hierarchically organized. The diagram illustrates the two-exon, one-intron repetitive make-up of the *B. mori* silk gene. Exon 2 illustrates the integrated and repetitive crystalline, noncrystalline, and border domains of the gene. It also shows that repeating arrays are variable in size. (Modified from Zhou et al. 2000.)

sequences (table 2.1; figure 2.1). *Ap-fhc* silks are composed of strings of poly-alanine (hereafter referred to as PAB, polyalanine block) with one of four, unique, nonpolyalanine blocks (hereafter referred to as NPAB, nonpolyalanine) (Sezutsu and Yukuhiro 2000). Some of the NPABs encode four repetitive motifs that show varying amounts of polymorphism. The variable and repetitive of amino acid sequences that make up the NPAB tails are the result of slippage and unequal pairing during crossover (Sezutsu and Yukuhiro 2000).

Spider *MA* silk is also composed of PABs and NPABs. The PAB blocks of *MA* silks sequenced to date have fewer alanine residues (4–10) than the sequence motifs that make up larval *Ap-fhc* silk (table 2.1). However, like *Ap-fhc*, a variable amino acid domain and a constant amino acid domain follow the polyalanine regions of both *Ma1* and *Ma2*. Analogous to *Ap-fhc*, the variable domain in spider *MA* silk is proceeded by repetitive sequence of nucleotides may result in a genetic hotspot (GGA,GGP) that initiates frequent rearrangements of the gene (Jeffreys et al. 1985; Sezutsu and Yukuhiro 2000). The number and organization of the potential NPAB motifs in *Ma1* and *Ma2* silk, however, cannot be determined

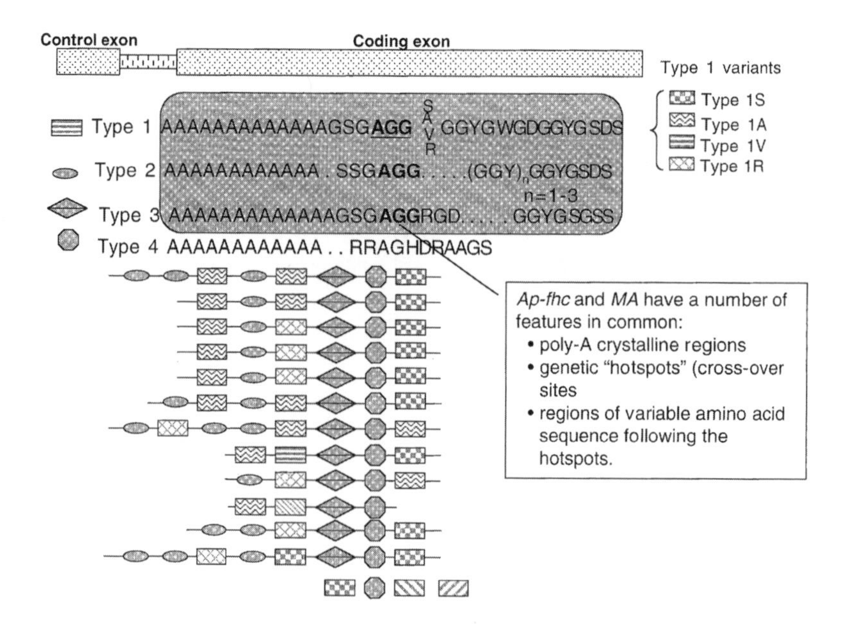

Figure 2.4 Repeating sequence structure of silks spun by *A. pernyi* is hierarchically organized. Genomic DNA for cocoon silks spun by *A. pernyi* show four motifs. The figure illustrates the constant and variable domains of the gene. The Chi-like triplet, AGG is underlined. Unlike *B. mori*, poly-alanine makes up the crystalline regions of the protein, and the noncrystalline regions are made up of a series of motifs which contain both constant and variable domains. (Modified from Sezutsu and Yukuhiro 2000.)

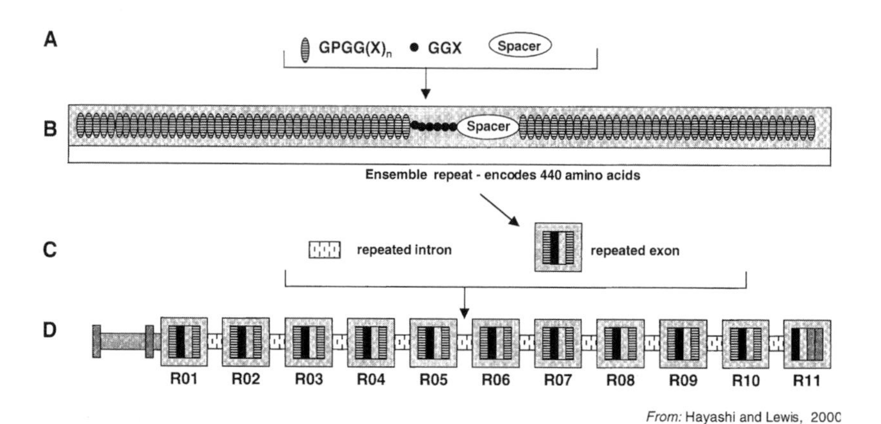

Figure 2.5 Repeating motif and structure of the *Flag* gene of *Nephila clavipes*. (A) Flag silk is also hierarchically organized. Repeat motifs GPGG(X)$_n$, and GGX are organized into an ensemble repeat. The ensemble repeats, in turn are organized into eleven repeating exons. Flag silk does not contain the GAGAX or poly-A sequences that encode the β-pleated sheet domains of the Fhc or MA silk. (B) Comparison of repeat motifs for the ten repetitive exons of two species, *N. clavipes* and *N. maculata* illustrates variation in repeat arrays. (Modified from Hayashi and Lewis 1998.)

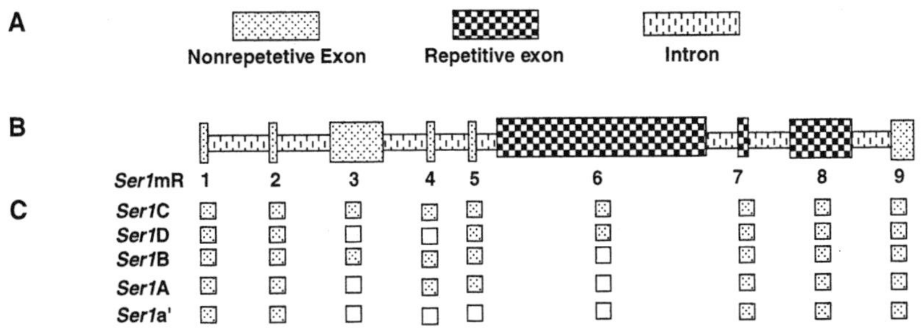

Figure 2.6 Repeating motif and schematic structure of the *Ser-1* gene of *B. mori.* (A) The schematic of the *Ser-1* gene show that the irregular size of the exons and introns as well as the positions of their amorphous and crystalline regions. (B) The primary sequence of (Ser1B) shows the amino acid sequences of exons 1–5 and 7–9. Unlike the *Fhc* genes, the crystalline and noncrystalline regions are not integrated in the Ser1B cDNA peptide. (Modified from Garel et al. 1997.)

due to the limited sequence data available. Nevertheless, the preliminary data suggest that the NPAB amino acid motifs of *Ma2* and *Ma1* differ in both sequence and size. In addition, although both contain polymorphic sites, *Ma2* silks are significantly more polymorphic than *Ma1* silks. The primary repeating motifs for *Ma2* silk are GPGXX/GPGQQ, GP(SYG), and A_n. The primary repeating motifs for *Ma1* are (GA), A_n, GGX.

Flag silks spun by *N. clavipes* do not contain PABs but do share the repetitive motifs GGX (alanine, serine) and GPGGX with the *Ma2* fibroin. Although some of the *Flag* sites are polymorphic, the sizes of each of the repetitive regions or exons of *Flag* silks are more similar to each other than to the size of the repetitive regions in either *MA* or *Fhc* silks. This suggests that *Flag* exons are either subject to homogenization (Hayashi and Lewis 2001) or that they are evolving under a more stringent selection than either *MA* or *Fhc.*

Fhc fibroin genes (and perhaps *MA* genes) are characterized by a similar molecular genetic architecture of two exons and one intron, but the organization and size of these units differ

Three different types of molecular genetic architecture have been identified for silk fibroins and their related fibrous proteins. These include the molecular genetic architecture of the *Fhc* silks (figures 2.3 and 2.4), of *Flag* silk (figure 2.5), and of the fibrous proteins such as *Ser1* (figure 2.6).

The *Bm-Fhc* gene is made up of two exons in which the hexapeptides are hierarchically organized (figure 2.3). Each encodes for both crystalline and noncrystalline protein domains that result in a 5263 amino acid polypeptide. *B. mori* exon 1 contains 25 bp of untranslated domain and a 42 bp region that encodes for 14 amino acids (Zhou et al. 2000). The repetitive domain of exon 2, separated from exon 1 by an intron of 970 bp, is organized into 12 subdomains that vary in size from 111 bp to 255 bp. The repetitive motif of exon 2 is similar to that of exon 1, but less well conserved. In addition to the repetitive domains are flanking, nonrepetitive amino acid regions that are highly conserved and that code for the noncrystalline components of the protein (Mita et al. 1994). Finally, exon 2 contains two different types of boundary or spacer units whose sequences are nonrepetitive.

The *A. pernyi-Fhc* gene is also made up of two exons and one intron, but the total length of the coding region is about one half that of *Bm-Fhc* (2.3 and 2.4). The full length of the *A. pernyi-Fhc* is about 8.1 kb, and it codes for a polypeptide chain of 2639 amino acids. Like *Bm-Fhc*, *Ap-Fhc* is characterized by one short exon encoding 14 amino acid residues, one intron (120 bp), and a second, larger exon encoding 2625 residues. Unlike the *Bm-Fhc* fibroin, *Ap-Fhc* contains a unique sequence of 155 residues of amino acid terminal sequence. The hierarchically organized units are predominately alanine (43%), glycine (27%), and serine (11%). The *Ap-Fhc* intron shows no nucleotide similarity to the *Bm-Fhc* intron, and exon 2 of *Ap-Fhc* and *Bm-Fhc* are also dissimilar as well (Tsujimoto and Suzuki 1979; Mita et al. 1994; Sezutsu and Yukuhiro 2000). The first exons of *Ap-Fhc* and *Bm-Fhc* are identical at 11 out of 14 residues (Tamura et al. 1987), which may indicate a similar function (Sezutsu and Yukuhiro 2000).

Exon 2 encoding of the *Ap-Fhc* protein contains 80 repetitive regions made up of linked polyalanine (PAB) and nonpolyalanine (NPAB) units (2.3). Seventy-eight of the 80 repetitive regions can be grouped into four types of motifs based on their NPAB sequences. Type 1 motifs are made up of four subtypes. Type 2 motifs differ in GGY triplets, suggesting that replication slippage events may be responsible for their generation. Type 3 motifs contain an Arg-Glycine-Asp (RGD) triplet. Type 4 is highly heterologous to the other motifs (Sezutsu and Yukuhiro 2000). *Ap-Fhc* includes variable domains that are not found in *Bm-Fhc* fibroin silks. Furthermore, the PAB sequences are highly conserved, but NPAB sequences are variable. This suggests either that the NPAB region of the molecule is evolving under less stringent selective factors than the PAB region or that the unique properties of fibroin result from positive selection on the NPAB sequences (Sezutsu and Yukuhiro 2000).

The molecular genetic architecture of the *MA* silks spun by *N. clavipes* has not been determined. However, based on 1951 base pairs of genomic transcript and cDNA identified as *Ma1* and *Ma2*, some investigators have proposed that the *Nc-MA1* silk protein is encoded by a single exon (Xu and Lewis 1990; Hinman and Lewis 1992; Beckwitt and Arcidiacono 1994; Guerette et al. 1996; Beckwitt et al. 1998). It may be that both *MA* proteins have an organization similar to *Ap-Fhc* (figure 2.2A,B).

The identified portions of *MA* sequence are hierarchically organized and contain characteristic, repetitive motifs that are common across ancestral and derived species (Gatsey et al. 2001). As in the *Fhc* genes, the number of amino acid, tandem repeats that make up *MA* is not rigorously conserved but retains a hierarchical and structural identity. This contrasts with the nearly exact repeats that characterize the *Bm-fhc* gene. Although it is not possible to analyze *MA* sequences in detail due to the relatively small portion of cDNA and gene that have been sequenced, the pieces identified so far are strikingly similar in organization to those observed in *Ap-fhc* gene (Sezutsu and Yukuhiro 2000). Like *Ap-fhc*, both *Ma2* and *Ma1* repetitive units are organized into poly-alanine blocks (6–10 residues) and nonpoly-alanine blocks (16–43 residues). Furthermore, like *Ap-fhc*, *Ma1* and *Ma2* seem to have ensemble repeats that are nonrandomly distributed across the cDNA transcripts sequenced to date. Finally, a potential genetic "hotspot," AGG, is located in a similar position and precedes the variable domains of both the spider and *A. pernyi* NPABs. Such similarities suggest that *MA* silk, similar to *Ap-fhc* fibroin, may also be subject to dynamic rearrangements.

The *Flag, Ser,* and *BR* genes are made up of multiple exons and introns

Flag, *Ser*, and *BR* genes, made up of multiple exons and introns, have a completely different genetic architecture from the *Fhc* and *MA* genes (figures 2.5 and 2.6). In the cases of *Ser* and *BR*, some of their exons encode for crystalline protein regions and others for noncrystalline protein regions. Therefore, primary coding regions of the protein glues are not homogeneous and selective splicing of mRNA transcripts results in proteins with different structural domains. In the case of the *Flag* gene, eleven of its thirteen exons are made up of similar repetitive units. None of the exons encodes β-pleated sheet structures, and instead encode protein domains that have been suggested to be a mix of β-helical and β-spiral structural domains (Hayashi and Lewis 1998). Selective splicing of *Flag* mRNA's exons would result in different sized protein domains that, in turn, could affect the protein's structure and function.

The *Flag* gene spans a total of 36 kb, of which 15.5 kb are protein coding (Hayashi and Lewis 2000). Hence, the size of the transcribed *Flag* and of *Bm-Fhc* proteins are similar, and may indicate a general functional or mechanical limit to fibroin size. None of the 13 exons that make up *Flag* contain poly-alanine and hence crystalline structural domains. Exons 3–12 encode the amino acid motifs GPGGX and GGX, which occur in tandem arrays of 440 amino acids with a nonrepetitive, highly conserved spacer unit (table 2.1). In the primary repeating motif GPGG$(X)_n$, alanine (32%), serine (26%), valine (11%), and tyrosine (20%) occupy 89% of the polymorphic sites. In the second primary repeat motif, alanine (46%), serine (39%), and tyrosine (11%) occupy 96% of the polymorphic sites of GGX. Thus, repeating motifs of the *Flag* gene contain more polymorphic sites than do the repeating motifs of the *Fhc* gene. *Flag* exons 1 and 2 encode for a nonrepetitive terminal region. Exon 13 codes for the 3′ end of the molecule, which includes both nonrepetitive and repetitive regions like those that characterize the previous exons (figure 2.5). Furthermore, the sizes of the *Flag* exons are much more homogeneous than the size of the repeating units of *Bm-Fhc*. The extreme similarity between some of the introns suggests homogenization due to recombination (Hayashi and Lewis 2000).

Ser1 is the best-known sericin gene, although partial sequences for at least four others have been identified (Grzelak 1995). *Ser1* is 23 kb long and is made up of nine exons and eight introns (figure 2.6). These include a large, central, alternative exon, which encodes for 60 repeats of a characteristic 114 bp motif (exon 6), in addition to two other exons that are repetitive (Garel et al. 1977; Michaille et al. 1986). Five different *Ser1* isoforms, each playing a different functional role in the cocoon, have been identified (figure 2.6; Zhou et al. 2000). As can be seen, the sericins are diverse.

There are five closely related *BR* genes, and they have been partially characterized. All of the *BR* genes have an exon/intron organization. Like the *Ser1* gene, *BR* exons code for either repetitive arrays or nonrepetitive arrays, but not both. While the *Ser1* gene encodes three differently sized, repetitive domains, the repetitive regions of *BR2*, *BR2.2*, and *BR6* encode four proteins that contain an uninterrupted block of about 100 tandemly arranged repeats organized into two repeat units (Paulsson et al. 1990). The architecture of the repetitive regions of the *BR3* gene is different, however. The repetitive array of the *BR3* is interrupted by 38 introns. The resulting exons are more variable in size and sequence than the repetitive arrays of the other genes. The presence of the introns in *BR3*, and the small sizes of some of its exons, may prevent exon homogenization that is dependent on unequal aligning of homologous sequences (Paulsson et al. 1990). In the case of *Flag*, homogeneous introns separate similar-sized exons. The extreme homogeneity of the size of *Flag* exons suggests that they are the product of a recent duplication event and/or that exon size is maintained by stringent selection. The evolution of *Br3* may be more complicated or selection for function may be relaxed.

The importance of the nine exon/intron structure of *Ser1* is that it allows multiple protein variants to be produced from a differentially spliced mRNA transcript. Specifically, differential splicing of *Ser1* gives rise to the five different types of mature mRNA transcripts (figure 2.6). In contrast, variation in *Fhc* silks is the result of unequal cross-over during recombination and largely limited to variation in allele length. Furthermore, new *Fhc* variants are produced during meiosis and hence the frequency of their appearance is limited by the length of insect generation time. However, selective splicing of the exons that encode *Ser* mRNA can generate multiple protein variants at different times during the larvae's development. Therefore, proteins with different structural configurations and properties can be produced from a single gene within the same individual at each transcription event. In the case of the *Ser* gene, the transcribed sericin proteins range from 65 to 400 kDa: exons 1 (93 bp, 13 codons), 2 (31 bp, 10 codons), 9 (569 bp and 369 bp with coding sequence followed by 188 bp of noncoding sequence), 7 (129 bp, repetitive), and 8 (114 bp repeats, repetitive) are constitutive and therefore found in all of the sericins. Exons 3 (1314 bp, 483 codons), 4 (93 bp, 31 codons), 5 (78 bp, 26 codons), and 6 (6.6 kb, repetitive) are alternative (Prudhomme et al. 1985). Exons 6 and 8 code for a 40% serine-rich motif and result in most of the β-sheet structure (Komatsu 1985). The peptides encoded by exons 3, 6, and 8 have specific physicochemical and structural properties that are responsible for the sliding and sticking of the sericins to the fibroin. Peptide 3 lends the property of high water solubility (Grzelak 1995).

Ser 2 is 16,000 bp and codes for two mRNA transcripts (5.4 bp and 3.1 kb) (Couble et al. 1987; Michaille et al. 1989). One is 3100 nucleotides, and the second is between 5000 and 6400 nucleotides, depending on which allele is considered (Garel et al. 1977; Michaille et al. 1986). The three other sericin genes, *MSGS-3*, *-4*, and *-5*, code for 3500, 2950, and 450 nucleotide mRNAs, but their structures are unknown (Grzelak 1995). The expression of both known genes (*Ser 1* and *Ser 2*) is spatially and developmentally regulated. Differential splicing of the primary transcript of *Ser1* is also developmentally regulated (Ishiwaka and Suzuki 1985; Couble et al. 1987; Michaille et al. 1989).

In summary, the multiple exon organization of the sericin gene and the lack of homogeneity among the exons results in considerable flexibility in the expressed proteins. In contrast, the exons that make up *Flag* are so similar that the effects of selective splicing of the transcript would largely limit variation to size differences of the transcribed regions.

Sequences coding for crystalline and noncrystalline protein domains are integrated in the repetitive regions of *Fhc* and *MA* exons, but not in the protein glues *Ser1* and *BR-1*

A result of the two-exon structure of *Fhc* and *MA* silks described above is that the primary regions coding for its crystalline and noncrystalline protein domains are integrated into one large, repetitive exon. In contrast, the crystalline and noncrystalline regions of the protein glues *Ser1* and *BR-1* are encoded by unique exons. According to the introns-first theory of gene evolution (genes with intron/exon structure evolved first; genes without introns are derived via intron loss), the exon/intron structure of the proteins *Flag*, *Ser*, and *BR* is more ancient than the structure of the *Fhc* and *MA* silk genes (Souza et al. 1998). However, phylogenetic evidence shows that the *Flag* gene is the most recently evolved type of silk produced by spiders. The amino acid sequences of the *Flag* protein are most similar to the repetitive units of *Ma2*. Comparison of the two proteins suggests that *Flag* may have evolved from *Ma2* through the loss of the crystalline, poly-A regions and subsequent gene duplication.

Genetic "hot-spots" promote recombination errors in *Fhc, MA,* and *Flag*

Glycine, alanine, and serine comprise about 53–97% of the amino acids of *Fhc* silks spun by the Bombycoidea, and 42–78% of *MA* silks spun by araneoid spiders (C.L. Craig, unpublished). In both cases, a high proportion of their repetitive amino acid and nucleotide motifs are made up of minisatellite sequences. Minisatellite sequences function as genetic hotspots, or recombination signals that induce crossover events. The instability they introduce can enable gene conversion (Lam et al. 1974; Jeffreys et al. 1985; Paulsson et al. 1992). However, gene instability may also generate adaptive variation. The ability of the minisatellite sequences to produce variants in the NPAB sequences of *Ap-fhc* can be viewed, in and of itself, as an evolved response to natural selection (Burch and Chao 2000).

Genetic hot spots are thought to result in the dynamic reorganizations of four types of repetitive arrays in the NPAB sequences in *Ap-fhc*. Even though there is high sequence conservation in length of the PABs of *Ap-Fhc* (PAB's sequence lengths in *MA* silk are less homogeneous), the noncrystalline portion of the NPABs is variable (figure 2.4; Sezutsu and Yukuhiro 2000). In the case of the *Ap-Fhc* gene, gene sequence rearrangements are attributed to the GCAGGUGGU nucleotides that result in the amino acid sequence AGG. AGG initiates all variable regions in the *Ap-fhc* motifs and in *Ay-fhc* as well as in *Antherea yamamai* fibroin silk (Sezutsu and Yukuhiro 2000). The

NPAB rearrangements include multiple duplication events as well as a triplication event of a 558-bp sequence. Similarly, although only limited data are available, the repetitive nucleotide sequence that encodes GGA in *Ma1* and PGG in *Ma2* silks spun by *N. clavipes* may contribute to the substantial allelic variation observed in these proteins that results from cross-over during recombination (Beckwitt et al. 1998). *Flag* silks, too, contain genetic hotspots. Some 75% of the protein is made up of proline, glycine, and alanine. The observed length differences in the tandem repeats of $GPGG(X)_n$ and GGX are probably due to cross-over events (Hayashi and Lewis 2000).

Codon bias, structural constraint, point mutations, and shortened coding arrays are alternative means of stabilizing precursor mRNA transcripts

Codon bias, structural constraint, point mutations, and shortened repetitive arrays may result in stabilization of the repetitive regions of silk proteins during transcription. The nucleotide biases in *Bm-fhc* codons differentiate the noncrystalline and crystalline regions in the protein molecule. In the case of *Bm-fhc*, the resulting codons determine the secondary structure of the mRNA (Mita et al. 1998; Nakamura et al. 1991). The first and second codon positions show a GC bias that is about 80% due to the presence of glycine (GGN) and alanine (GCN) residues (Zhou et al. 2000). Zhou et al. hypothesized that the alternate dicodon, GGU-GCU for GlyAla repeats, results in an inverted stem-loop structure in the crystalline subdomains of the mRNA transcript. The resulting structural constraints, in turn, maintain the secondary structure of the fibroin mRNA (Mita et al. 1988). No stem-loop structures are formed in the boundary elements due to high codon diversity. Stem structures are only weakly formed in the region of the transcript coding for amorphous domains (Zhou et al. 2000).

In contrast to GCU codon bias for alanine in the *Fhc* gene, the most abundant codon for alanine in *Ap-fhc* is GCA. Polyalanine in *Ap-Fhc* occurs in blocks of 13 residues. The GCA codon for alanine occurs in 662/1137 times, or across 58% of the coding region (Nakamura et al. 1991; Sezutsu and Yukuhiro 2000). The maximum number of GCA isocodons in each PAB is seven. Therefore, the tri-nucleotide repeat sequences that encode poly-A could be subject to frequent replication slippage events (Sezutsu and Yukuhiro 2000). Nevertheless, the variation in alanine number is small and may indicate significant constraint on PAB size. Stable alleles that retain tri-nucleotide repeats are often interrupted by point mutations that, as in the case of *Ap-fhc*, may stabilize the repeat tract and limit the size of the polyalanine region (Sezutsu and Yukuhiro 2000).

Like *Fhc*, all of the spider silk genes sequenced to date show an extreme A/T bias (80–90%) in third position nucleotides and a preference for G/C in the first two codon positions (Xu and Lewis 1990; Hayashi and Lewis 1998). The polyalanine regions of the *MA* silks, however, are variable (4–10 alanine) and smaller than the PAB in *Ap-fhc*. Data on the possible PAB of *MA* silk are limited and based on cDNA alone. Therefore, one can only speculate that the poly-alanine sequences of *MA* are subject to less stringent functional constraint than are the PAB regions of *Ap-Fhc*. Alternatively, data on the entire *MA* gene and protein could reveal a larger order of organization that differs from that of *Ap-fhc*. We need to know the complete sequences of the *MA* fibroin gene as well as the 5′ flanking sequences (Sezutsu and Yukuhiro 2000).

The coding regions of the *Ma2* gene and *Flag* exons contain multiple polymorphic sites, more than are found in the *Ma1* transcripts. Point mutations in the repetitive sequences of *Flag*, like point mutations in the polyalanine sequences in NPAB, may stabilize DNA during recombination and may also stabilize the mRNA transcript during transcription.

Differential regulation of gene expression and selective splicing may allow rapid adaptation of silk functional properties to different environments

Evolvability—the generation of variation—is a property of the genotype and distinct from the phenomenon of variation among individuals in populations (Wagner and Altenberg 1996; Kirschner and Gerhard 1998). Silk proteins are highly evolvable, as defined above, due to their highly repetitive amino acid and nucleotide sequences and the presence of apparent genetic hotspots. Despite their instability, there is extreme conservation and convergence in spider fibroin sequences suggesting that they are evolving under considerable functional constraint and conflict (Gatsey et al. 2001).

Orb-spinning spiders forage in all but the most extreme terrestrial habitats. If all silks were made up of only one protein, then new functional variants of the silk could result only from the slow accumulation of genetic differences across spider generations. Alternatively, because some spider silks are made up of more than one protein or characterized by an intron/exon molecular architecture, differential expression of protein type or selective splicing of mRNA transcripts could result in the rapid adaptation of silk proteins to new environments.

Recent field data seem to suggest that adult araneoid spiders may be able to adapt the composition of *MA* silk to the physical environments in which they forage through differential protein expression

(Craig et al. 2000). For example, *Ma1* cDNA is largely composed of $(GA)_n$ and $(A)_n$ and a GGX motif. The alanine units, as discussed above, result in the crystalline, β-sheets. The repeated glycine motif and variable site is thought to form a 3_{10}-helix that links the crystalline regions together. Therefore, the *Ma1* mRNA transcript is largely responsible for silk strength. *Ma2* contains the repeating amino acid motifs GPGXX/GPGQQ, GP(SYG) (table 2.1). A recent model for *Ma2* silks proposes that these motifs result in β-turns that, acting like springs, result in fiber elasticity. Cohesion of the spring-like structures is dependent on hydrogen bonding networks that form between spirals at their variable sites (Hinman and Lewis 1992; Hayashi and Lewis 1998). Because silk elasticity is attributed to the hydration of the *Ma2* encoded protein, the differential expression of *Ma1* and *Ma2* in different humidity environments could result in silks with similar elasticity and strength despite the fact that they are produced and function in different relative humidities. In the extreme, *MA* silk made up solely of *Ma1* transcript would be strong but inelastic and silks made up solely of the *Ma2* transcript would be less strong but elastic. The flexibility afforded by the two-protein composition of spider *MA* silk as well as the exon/intron organization of *Flag* silk may be an important correlate of the burst of speciation that resulted with the emergence and speciation of the Araneoidea.

One way to test the prediction of environmental flexibility in silk gene expression would be to compare the amino acid composition of the *MA* silks collected from spiders foraging in different sites and in particular to compare the relative proportions of serine, glycine, and proline residues in the total amino acid content of a silk. For example, the *Ma2* silk transcript for *N. clavipes* that has been sequenced to date contains 57 serine residues to 176 glycine residue (32%). *Ma1* silks contain 34 serine residues to 307 glycine residues (11%). Therefore, *MA* silk will contain almost five times more glycine to serine if *Ma2* and *Ma1* protein transcripts are produced in equal portions. Furthermore, the amount of proline present in an amino acid analysis would be linearly proportional to the presence of the *Ma2* transcript.

Summary

The preceding chapter examines the three types of molecular genetic organization of silk genes that have been found. The primitive Mesothelae and the Mygalomorphae produce silks that are fibrous, protein glues; according to a phylogenetic hypothesis, protein glues evolved first in spiders. The structure of the spider genes that encode protein glues is unknown, but may be similar to that of the *Ser1* gene of *B. mori* and the Balbiani Ring (*BR*) genes of *Chironomus tentans* that encode protein glues produced by insects. *Ser1* and *BR-1* genes contain exons that encode only for crystal-forming regions and exons that encode only for noncrystalline material.

The second type of molecular genetic organization characterizes the gene for cocoon or fibroin silks produced by Lepidoptera larvae and probably *MA* or the dragline silk produced by spiders. The gene is made up of only two exons and one intron. The first exon is thought to be primarily regulatory in function, the second exon encodes for both crystalline and noncrystalline domains of a silk fibroin. This second exon is made up of repetitive elements that are hierarchically organized into crystalline and noncrystalline regions. The compositions of the repetitive units are similar but their sizes are heterogeneous.

The third type of molecular genetic organization characterizes the *Flag* silk genes. Like *Fhc* and *MA* silk genes, *Flag* is hierarchically organized. Unlike *Fhc* and *MA*, *Flag* is made up of multiple exons and introns. The introns are nonrepetitive and highly similar to each other. The exons are repetitive, similar to each other, but different from the introns. The *Flag* exons do not contain the polyalanine, crystalline domains that make up the *Ap-Fhc* and *MA* silks. Nor do they contain the GAGAG(S) and GAGAGX regions that make up the crystalline regions of the *Bm-Fhc* silks. Furthermore, the exon/intron organization of *Flag* genes should allow greater opportunity for evolutionary innovation through selective splicing of the mRNA transcript.

3

The Mechanical Functions of Silks and Their Correlated Structural Properties

with C. Riekel

Chapter 1 provides background information on why and how different amino acids can affect the folding patterns of silk proteins. In considering these configurations, chapter 2 describes how molecular genetic processes can result in the substitution of one amino acid for another. The remaining chapters identify factors that can act as selective agents on the protein variation that the molecular genetic processes generate.

Five hypotheses have been advanced to explain the diversity and evolution of silks spun by the Orbiculariae: (1) random genetic events (Rudall and Kenchington 1971); (2) selection for synthetic efficiency (Mita et al. 1988; Candelas et al. 1990; Hayashi and Lewis 1998; chapter 2); (3) selection for mechanical properties such as strength and elasticity (Denny 1976; Denny 1980; Gosline et al. 1984; Craig 1987a; Gosline et al. 1994; Köhler and Vollrath 1995); (4) selection for reduced reflectance and hence reduced web visibility (chapter 4); and (5) selection for amino acids that may reflect the composition or diet of the silk producer's prey (chapter 7). None of these hypotheses is exclusive of the others as each addresses a different aspect of silk synthesis or function. For example, the three different types of fibroins that spiders spin are made of highly repetitive sequences of amino acids that result in a high probability of cross-over, random deletions, or duplications of DNA segments during recombination (Hypothesis 1). While selection may act to constrain variation or homogenize the silks, errors during replication could also result in divergent protein compositions that effect positive selection for silks with new properties (Hypotheses 2, 3, and 4). Different physical properties of silks translate into different mechanical and reflectance properties of webs. These in turn may bias the prey that webs intercept and hence the pool of amino acids available to the spider (Hypothesis 5).

Hypotheses 3 and 4 suggest that the primary selective force on evolving silks acts on their unique functional properties. The physical properties of silk—strength, elasticity, and flexibility—are determined by the structural organization of the protein and influenced by its amino acid composition. Silks spun to intercept fast-flying prey need to be elastic and strong, to withstand insect impact, as well as difficult to see, to prohibit the insects from seeing and avoiding them (see chapters 4, 5, and 6). Silks spun to protect spider's eggs need to be hydrophobic and tough; silks used to wrap prey need to be flexible; and silks used to decorate webs need to be highly reflective if they are to attract prey to them. In this chapter, the molecular organization of silks in light of their functional properties will be reviewed.

Fibrous proteins produced by primitive spiders are not spun into weight-bearing threads, such as the dragline silks produced by the araneomorph spiders, nor do primitive spiders build free-standing webs that can intercept flying prey. Instead, nonaraneomorph spiders use fibrous proteins to line underground burrows and to bind leaves, soil, and rocks into protective shelters. Virtually all of the nonaraneomorph spiders (Liphistiomorphae, Mygalomorphae) construct burrows and catch prey at the entrance of a protected retreat (figure 3.1). Secondarily, some spiders expand the area over which prey can be detected by spinning "trip lines" that radiate away from the entrance of their retreats. The expanded use of trip lines is thought to have led to the evolution of the dense masses of silks in which prey are entangled. These silks can extend the area over which spiders can sense, trap or detain the insects that chance into them (Coyle 1987).

Coyle's data (1987) suggest that (1) foraging from an underground burrow is the ancestral foraging mode of spiders; (2) silks first evolved as a mortar used to bind soils and vegetation; and (3) sticky fibrous proteins were adapted to above-ground foraging sheets. The data are plotted on two, recent phylogenies and analyzed using the parsimony option in MacClade (Maddison and Maddison 1992; figure 3.2). Information on the ancestral character state for all families is not available, and therefore all character states were included in the analysis. Phylogeny 1 is based on Goloboff (1993) and phylogeny 2 is

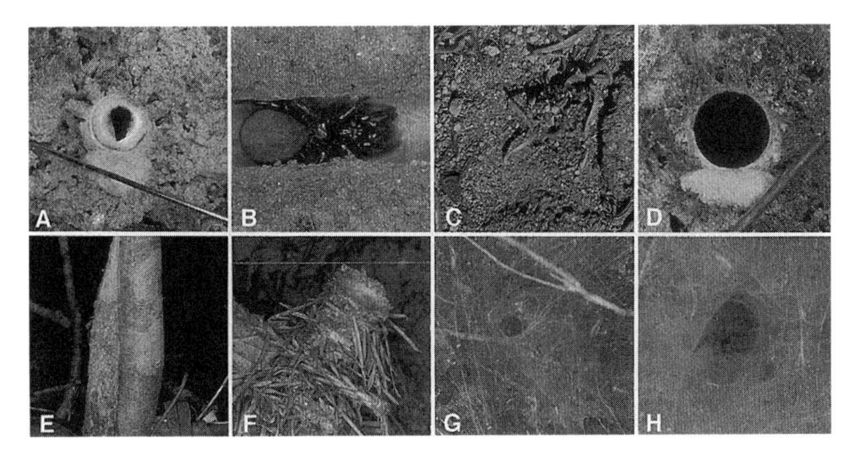

Figure 3.1 Nonaraneomorph spiders use fibrous proteins to construct protective burrows and retreats. (A) *Aliatypus aquilonius* (Antrodiaetidae), burrow opened to reveal sealing silk, Humbolt Co., CA. (B) *A. gulosus* in burrow, Orange Co., CA. (C) *A. aquilonius*, undisturbed entrance to burrow, Humbolt Co., CA. (D) *A. aquilonius*, open burrow entrance, Humbolt Co., CA. (E) *Sphodros atlanticus* (Aytypoidae), funnel web on tree trunk; Webster, NC. (F) *Atypoides riversi*, turret, Santa Cruz Co., CA. (G) *Telechoris striatipes* sp. (Dipluridae) web, Tsavo W, Kitani, Kenya. (H) *T. striatipes* in mouth of retreat, Kilifi, Kenya. (Photographs used with permission of F.A. Coyle.)

based on Raven (Raven 1985); they differ in the placement of the families Cyrtaucheniidae, Migidae, and Actinopodidae.

The data traced on phylogeny 1 (1 reconstruction) and phylogeny 2 (1 reconstruction), show that foraging from an underground burrow is the ancestral foraging mode for all spiders. With the exception of the Liphistiidae (Mesothelae) and Atypidae, all of the families of nonaraneomorph spiders include some groups that forage above ground. Furthermore, in both phylogenies, the use of silk to produce an above ground "burrow" is derived. Spiders that build above-ground funnels extend at least some silk away from the funnel entrance. While the data are not complete by any means, the shift from below-ground to above-ground foraging correlates, in both cases, with an order of magnitude increase in species number (J.E. Bond and M. Hedin, unpublished).

The derived use of fibrous proteins to produce a protective, above-ground burrow must have resulted in a substantial change in how selection affected protein properties. Above-ground burrows require at least two types of silks or silks with multiple properties: fibrous proteins (primitively used in conjunction with soil and vegetation) used as mortar, and fibrous proteins used to construct a funnel (figure 3.1). The differences between mortar and funnel silks have not been studied, but the structural data below suggest that all

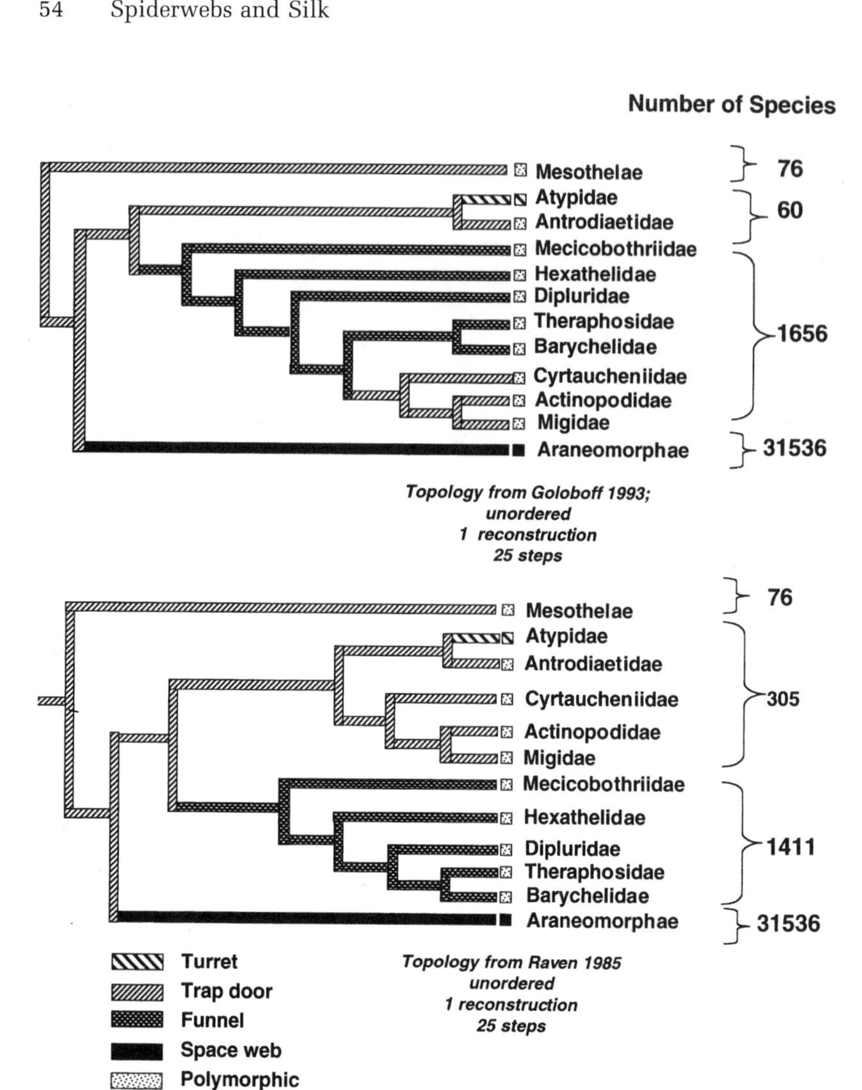

Figure 3.2 Character tracing of foraging modes on two phylogenies of nonaraneomorph spiders. (A) Phylogeny abstracted from Goloboff (1993). (B) Phylogeny abstracted from Raven (1985). Both phylogenies are based on parsimony analysis and no statistical support values for branches are available. The Liphistiadae (in the Mesothelae) and the Atypidae are the only families of nonaraneomorph spiders whose members forage exclusively from an underground burrow. Correlated with the divergence of the two primary groups of the Mygalmorphae, according to Raven, is the evolution of the above-ground funnel. Spiders that forage in above-ground funnels sit at the funnel entrance, suggesting that funnels represent the transposition of an underground burrow to an above-ground burrow. Number of species from Bond and Opell (1998); Raven (1985); and Platnick (2000). Foraging states not shown, but included in the analysis are collar, long tubular extension, rigid prey-sensing attachments, silk signal lines, and sheet. (Data from Coyle 1987.)

fibrous proteins contain crystals that are less organized than those in silk fibroins. The potentially new types of silk used for above-ground foraging may be the result of fiber manipulation or spinning behavior, as well as new silk processing systems. Above-ground silk would be under different selection pressures than those used below the ground. For instance, in the absence of a burrow, silks might be selected to be hydophobic and tough to enable sustained protection for the spider.

Nonaraneomorph spiders are not capable of substrate-free spinning but are dependent on their physical environment to pull silks from the spider's spinneret. As a result, the silks they produce are intimately associated with the surfaces on which they walk. If an insect becomes entangled in a nonaraneomorph web, it struggles against a sticky mass that is supported by a solid surface (the ground). With respect to foraging, the primary selection pressure on nonaraneomorph fibrous proteins and silks is to retain prey until the spider can subdue them. Unlike the araneoid webs, capture probability is not dependent on the ability of the silk to absorb insect impact (as is apparent for an aerial web) but on how well the short fibers that make up the sheets adhere to each other as well as to the insect. Web are extremely sticky sheets that, at the least, slow insects that walk onto them. The lack of web organization, in conjunction with the importance of prey retention, suggests that the selection pressures on the mechanical design of the nonaraneomorph silk sheets are less stringent (or at least different) than selection pressures on the webs spun by aerial web spinners. Ancestral spiders spin sticky sheets; derived spiders spin webs that are both sticky and energy-absorbing.

Ancestral araneomorph spiders spin "dry" capture silks into irregular webs; derived araneomorph spiders (here, the Orbiculariae) spin dry and "wet" capture silks into symmetrical webs

The evolution of an irregular aerial web and the evolution of cribellate silk (brushed silk that entangles prey) correlate with the phyletic divergence of the Araneomorphae, or true spiders, from the nonaraneomorphs (figure 3.3). The araneomorph spiders that spin cribellate silk manufacture a variety of webs. The cribellate orb weavers in the family Uloboridae (with the exception of *Miagrammopes* which have paracribellar silks) spin capture threads made up of two axial fibers surrounded by masses of tiny silk fibrils (figure 3.4; Peters 1983; Peters 1984; Peters 1986; Kovoor 1987; Eberhard 1988; Opell 1990; Peters 1992; Eberhard and

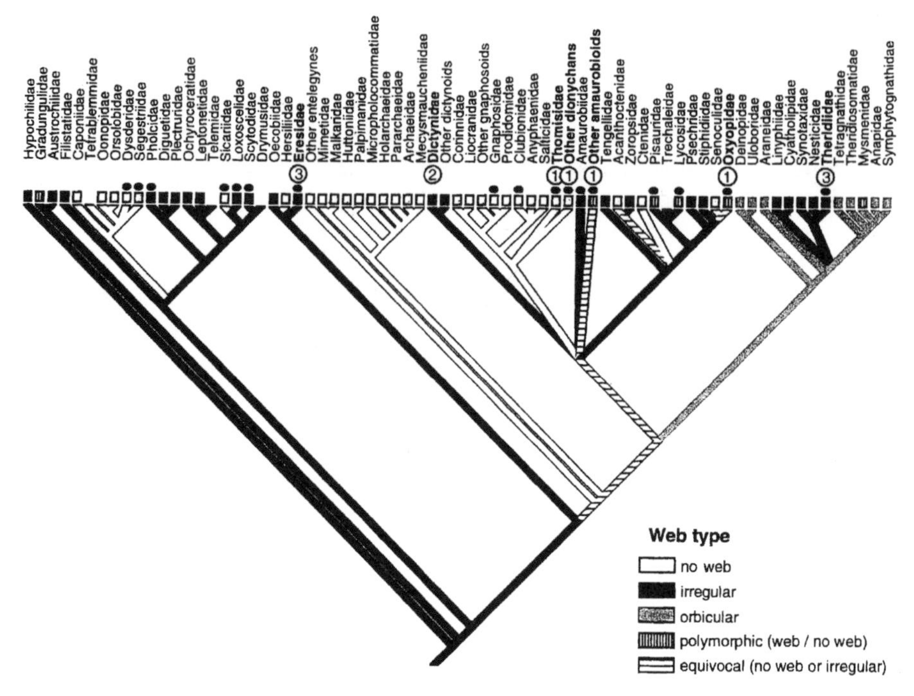

Figure 3.3 Types of webs spun by the araneomorph spiders. Foraging above ground and via and irregular web is the ancestral character state for foraging mode of the araneomorph spiders. Irregular webs probably evolved two times, but the orb web evolved only once. (Phylogeny based on parsimony; no statistical support values for branches are available. Personal communication with J. Coddington. Figure reprinted from Avilés 1997 with permission from Cambridge University Press.)

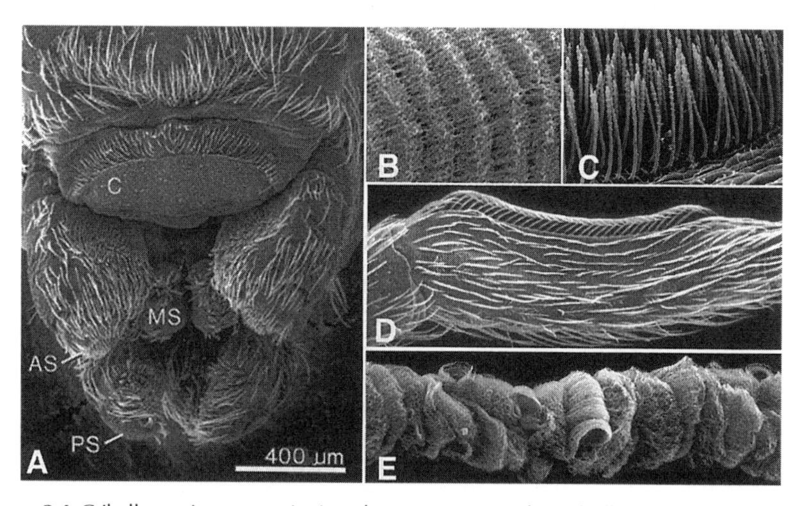

Figure 3.4 Cribellate spinnerets, spinning plates, spigots, comb, and silks. (A) Spinning apparatus of the cribellate spider *Zosis geniculatus*: C = cribellum; MS = median spinnerets; AS = anterior spinnerets; PS = posterior spinnerets. (Reprinted from Opell 1993 with permission from J. Wiley Co.) (B) Cribellar fibrils of *Matachia livor* (Opell 1999). (Reprinted from Opell 1999 with permission from Blackwell Press). (C) Cribellar spigots of *M. livor* (op cit.). (D) Calamistrum (comb to brush cribellate silks) of *Uloborus glomosus*. (Reprinted from Opell 2001 with permission from the author). (E) Cribellar thread of *Hyptiotes cavatus*. (Reprinted from Opell 1994 with permission from J. Wiley Co.)

Pereira 1993; Opell 1993; Opell 1994a; Opell 1994b; Opell 1995; Opell 1996; Opell 1999a; Opell 1999b). The threads are dry (as compared to the wet, gluey capture threads of derived species), and prey are retained via entanglement. Cribellate silks function analogously to the action of the coiled threads of a Velcro fastener but can also adhere to smooth surfaces (Opell 1994c). Insects that land or walk over the surface of cribellate webs come in contact with thousands of tiny fibrils that stick to the insect's body, adhering to surface irregularities and setae. Foraging via an irregular, above-ground web is the ancestral character state for foraging modes of all the araneomorph spiders. While foraging via webs has been lost, irregular webs probably evolved independently at least two more times (Avilés 1997). The orb web evolved only once and has been lost at least three times (*Hyptiotes-Miagrammopes* lineage; *Mastophora* lineage and theridiid-linyphiid lineage).

Spiders in twenty-two araneomorph families produce cribellate capture threads that entangle prey (Griswold et al. 1998). The axial fiber of catching silks spun by ancestral araneomorph web spinners is drawn from the ampullate gland (Kovoor 1987). In contrast, the axial fiber of catching silks spun by the derived cribellate spiders, Deinopoidea is drawn from a unique gland, the pseudoflagelliform gland (Kovoor 1987). The fibrils that surround the capture threads of both ancestral and derived cribellate species originate from the cribellar glands. The cribellar glands are the smallest silk glands known in spiders. They are typically acinous, or grape-shaped. In some species, they number in the thousands (Kovoor 1987). Each tiny gland feeds into a tiny spigot (figure 3.4). The silks are brushed en masse around the two, larger, axial threads via a tarsal comb (the calamistrum), also unique to the cribellates. The breaking strength of the entangling threads may be similar to the breaking strength of the axial threads and proportional to fibril diameter (Opell and Bond 2001; figure 3.5).

The Aranoeidea, the derived araneomorph spiders, do not have a cribellum. Their webs retain insects via entanglement as well as sticky glues that coat the capture thread (figure 3.6). The orb web is ancestral for all of the Orbiculariae and may have evolved only once. A viscid catching spiral, however, is unique to the araneoids (figure 3.7). Comparison of the stickiness of the cribellar entangling fibrils versus glue-coated threads shows that araneoids can spin stickier webs at greater material economy than the cribellate spiders (Opell 1997b; Opell 1998; Opell 1999a; Opell 1999b).

The composition of the glues the araneoids produce is variable and complex. The glue droplets are composed of a variety of amino acids and other low molecular-weight compounds that include inorganic ions, phosphorylated glycoprotein, lipids, and organic, low molecular-weight compounds (Tillinghast et al. 1981; Vollrath et

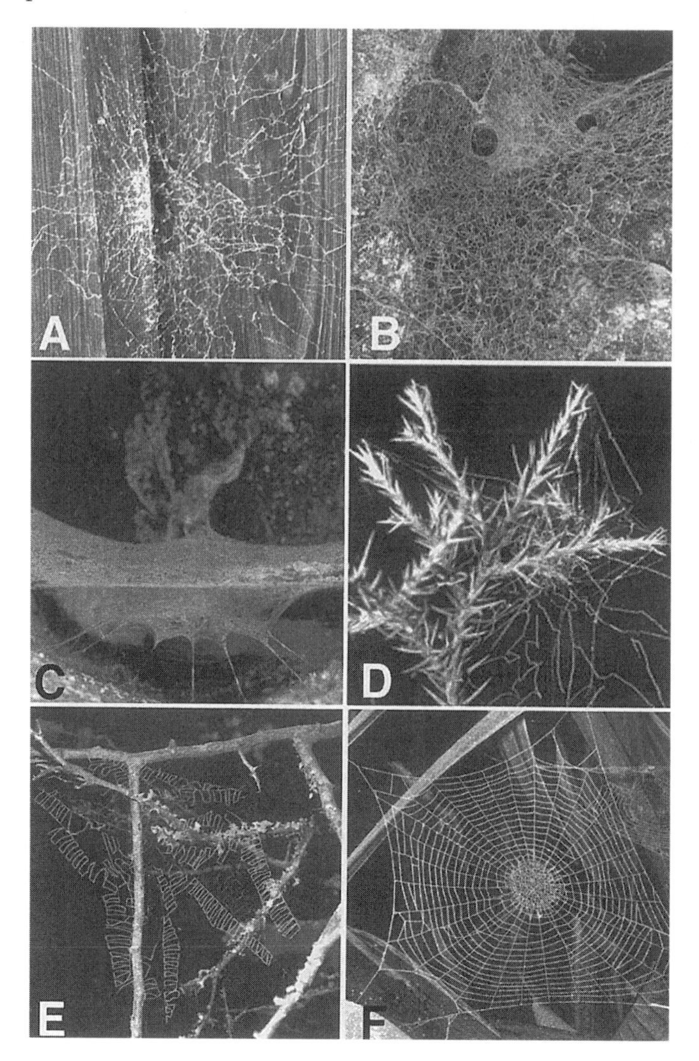

Figure 3.5 Some of the diversity of webs spun by cribellate web spinners. (A) *Kukulcania hibernalis,* radiating web. (B) *Badumna longinqua,* funnel web. (C) *Neolana pallida,* suspended sheet-web. (D) *Mexitilia trivittata,* cob-web. (E) *Matachia livor,* ray-web. (F) *Waitkera waitakerensis,* orb web. Webs A–E are spun by araneomorph spiders; web E is spun by an orb-spinning, araneomorph spider in the family Uloboridae. (From Opell 1999; reprinted with permission of *Biological Journal of the Linnean Society,* Blackwell Press.)

al. 1990; Vollrath and Tillinghast 1991). It may be that these data suggest that spiders may vary droplet composition, and hence functional properties of the viscid material, with changes in diet and/or changes in the physical environment in which the spiders forage (Higgins et al. 2001).

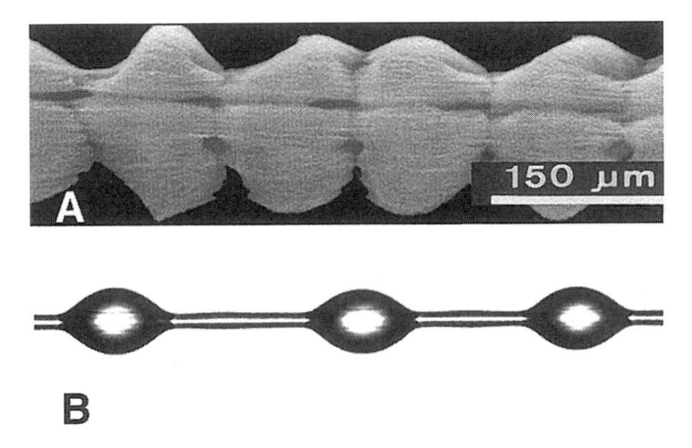

Figure 3.6. In contrast to the Deinopoidea, the Araneoidea produce a catching thread covered with viscid droplets. Prey adhere to the web when the glue smears across insect body surfaces at interception. (From Opell 1998 with permission from the author.)

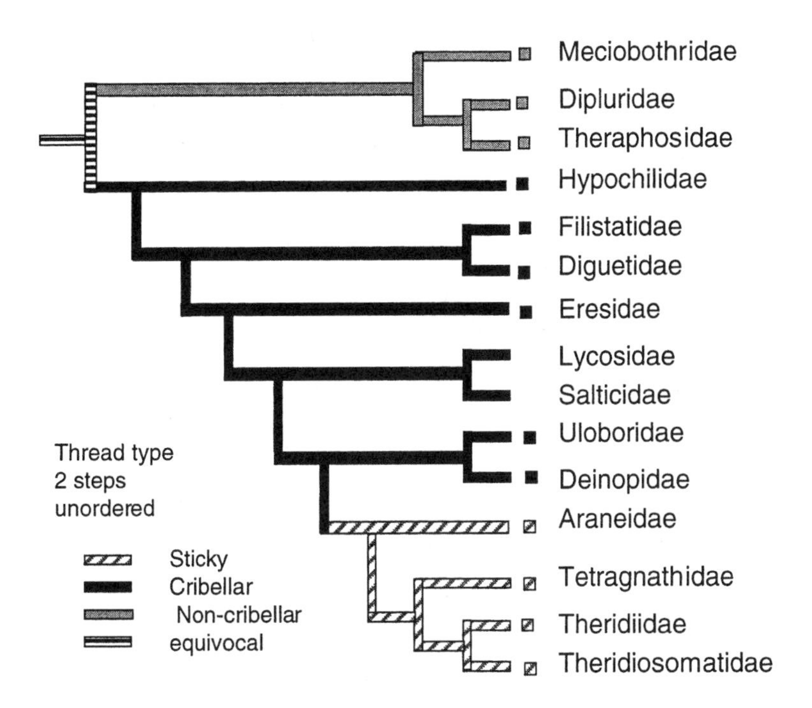

Figure 3.7 Character optimization for catching silk. The evolution of the flagelliform gland and a sticky capture thread correlates with the appearance of the araneoid orb-weavers and their diversification (Araneidae–Theridiosomatidae; redrawn from Bond and Opell 1998.) Phylogeny based on parsimony; no branch support values are available (pers. commun. with J. Coddington).

The Orbiculariae spin nets that are suspended under tension and that approximate minimum volume architectures

Approximately 4200 species of spiders spin orb webs or webs derived from an orb design (figures 3.8–3.12). The Oribiculariae include the primitive orb weavers (Deinopoidea) and "modern" orb weavers (Araneoidea) that diverged from other araneomorphs in the early Cretaceous (Wiehle 1931; Kovoor 1977; Coddington 1986; Seldon 1989; Coddington 1990a; Coddington and Levi 1991). The ability to produce an orb web, a symmetrically arranged, radial network of threads held under tension, represents an ancestral character state (a feature of primitive taxa) of the Orbiculareae and evolved prior to deinopoid and araneoid divergence (Coddington 1986; but see Eberhard 1990; Hausdorf 1999). Due to its free suspen-

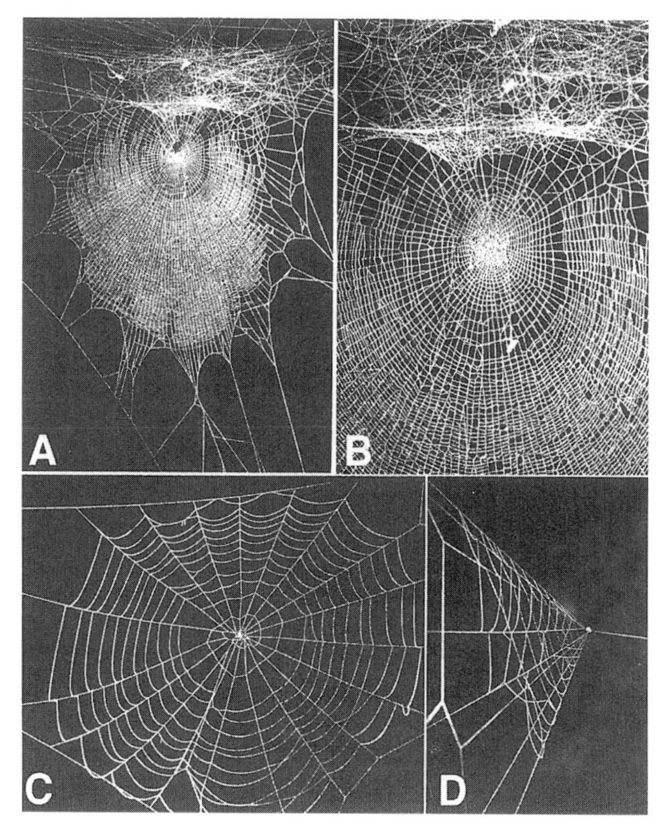

Figure 3.8 Webs of Araneoidea. (A) *Nephila clavipes* Tetragnathidae, La Selva, Costa Rica. JC. (B) Ditto, close-up of hub and retreat. JC. (C) *Bertrana laselva*, Araneidae. La Selva, Costa Rica. CG. (D) *Theridiosoma* sp., Theridiosomatidae. JC. Puerto Rico. Photographs of webs by J. Coddington, JC, and C. Griswold, CG. (From Griswold et al. 1998; reprinted with permission of *Zoological Journal of the Linnean Society*, Blackwell Press.)

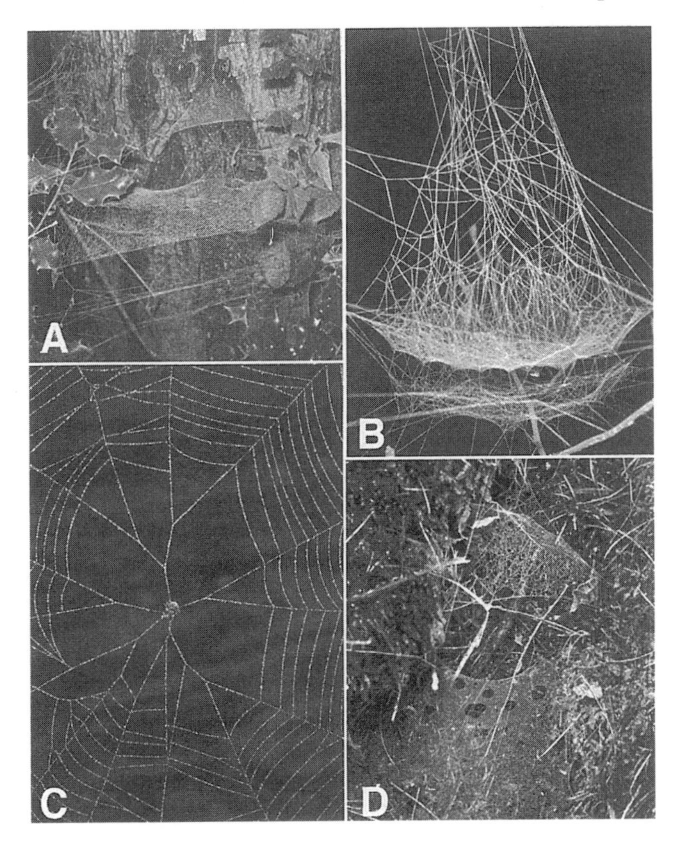

Figure 3.9 Webs of Araneoidea. (A) *Pimoa breviata* Pimoidae, Brookings, Oregon, USA. GH (B) *Fontinella pyramitela*, Linyphiidae, Patuxent, Maryland, USA. GH. (C) *Epilineutes globosus* (Theridiosomatidae), Barro Colorado Island, Panama. (D) *Tekella unisetosa*, Cyatholipidae, Fiordland, New Zealand, TM. Photographs of webs by G. Hormiga, GH, and Teresa Meikle, TM. (From Griswold et al. 1998; reprinted with permission of *Zoological Journal of the Linnean Society*, Blackwell Press.)

sion, the Orbiculariae can forage in new habitats not available to other nonflying predators: open space. Prey that are intercepted either do not see the web or are attracted to the web area by the reflectance properties of the silks. These new aspects of prey capture mean that insect visual capacity can become a selection factor on the evolution of orb webs, as may the light environment in which the web is suspended (chapters 4, 5, and 6).

The tensile webs built by the Orbiculariae approximate minimum volume architectures. A structure is a minimum volume design if: (1) every member of the structure is built of the same material and under uniform tension; and (2) if the stress (force/area) is equal to the breaking stress of the material. As a result of these specifications, the structure contains the minimum amount of material needed to withstand a specified set of loads. Spiders' webs are minimum volume

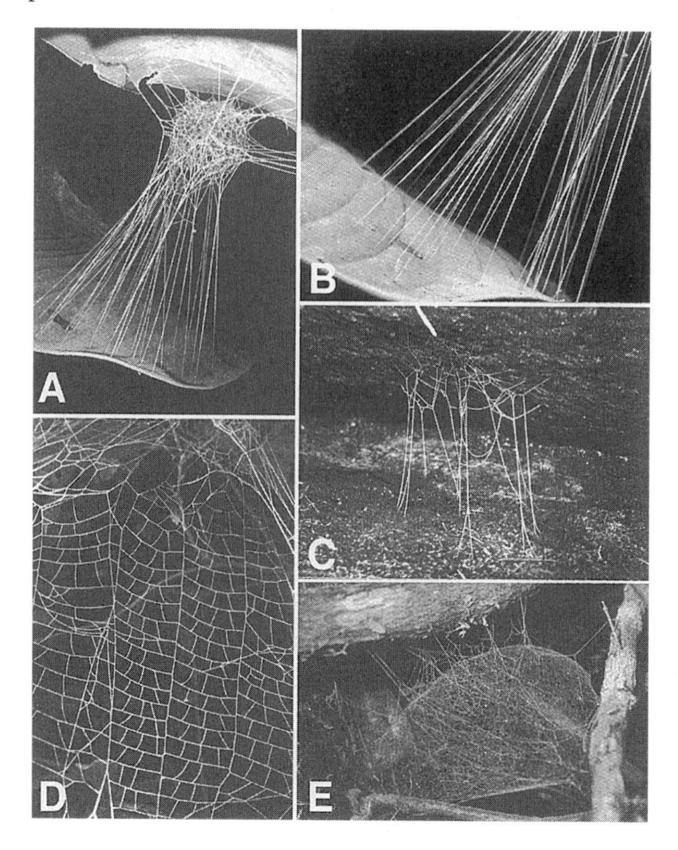

Figure 3.10 Webs of Araneoidea. (A) *Achaearanea* sp. Theridiidae, La Selva, Costa Rica. GH (B) Ditto, close-up of sticky gum-foot strands. GH (C) *Nesticus* sp., Highlands North Carolina, USA GH (D) *Synotaxus turbinatus*, Synotaxidae, La Selva, Costa Rica, capture lines and retreat. GH (E) *Pahoroides ichngarei*, Waippotta, New Zealand, TM. Photographs of webs by G. Hormiga, GH, and T. Meikle, TM. (From Griswold et al. 1998; reprinted with permission of *Zoological Journal of the Linnean Society*, Blackwell Press.)

designs when spun. Functionally, however, they can only approximate minimum volume architectures because almost any insect that is intercepted at a web damages it. The resulting distribution of forces may become biased.

The orb web is a key innovation and correlates with the emergence of the Orbiculareae spiders. In addition, the evolution of the Araneoidea (derived Orbiculareae), correlates with a transition from a horizontal orb that intercepts and retains prey via fiber strength, to a vertical orb that intercepts prey via fiber strength, extension, and elasticity, and that retains prey via viscid glues (Opell and Bond 2001). The glues are produced in the spider's aggregate glands and are secreted along with the flagelliform proteins (spigots for both located on posterolateral spinnerets; figure 3.6). While the Deinopoidea rely on the high breaking strength of cribellar threads

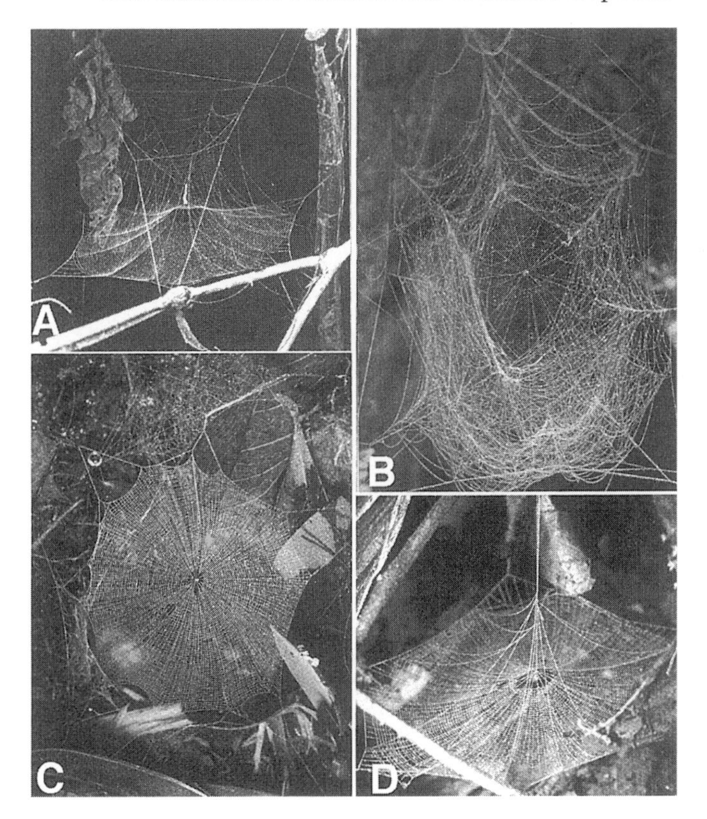

Figure 3.11 Webs of Araneoidea. (A) *Maymena* sp., Mysmenidae, La Selva, Costa Rica. (B) *Mysmena* sp., Mysmenidae, Cerro, Costa Rica. (C) *Anapistula* sp., Symphytognathidae, Yunque, Puerto Rico. (D) *Anapisona simoni* Anapidae, Llorona, Costa Rica. All photographs by J. Coddington. (From Griswold et al. 1998; reprinted with permission of *Zoological Journal of the Linnean Society*, Blackwell Press.)

to retain prey, the Araneoidea depend on the shear thinning properties of viscid glues and fiber extension and insect entanglement. In the case of araneoid webs, prey retention is no longer a function of the mechanical properties of the silk threads but a function of the viscous glues that coat the web spiral and possible aerodynamic dampening due to the silk's extensibility (Lin et al. 1995). As a result, the ability of the web to intercept and retain prey can become independent foci of natural selection.

Web function is determined by the interaction between web architecture and the material properties of silks

One of the most important architectural features of webs affecting function is the pre-stress at which the web is suspended (figure 3.13).

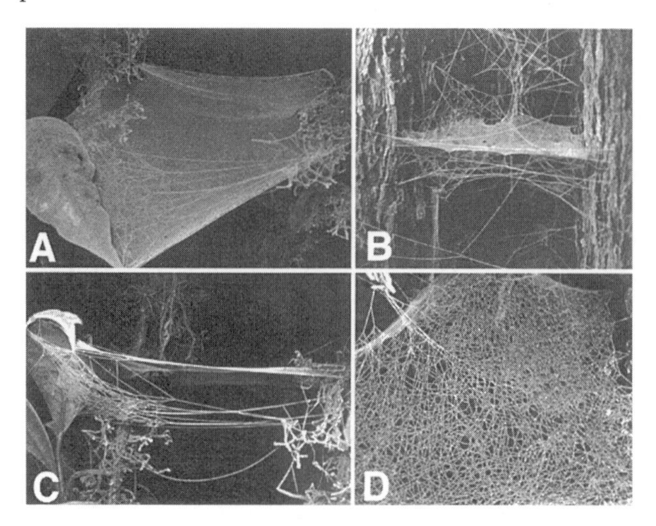

Figure 3.12 Webs of Araneoidea. (A) *Isicabu* sp., Cyatholipidae, Mt. Cameroon, Cameroon GH (B) *Teemenaarus* sp. Kuranda, Australia. GH (C) *Isicabu* sp., Cyatholipidae, Mt. Cameroon, Cameroon G. Hormiga (D) *Frontinella pyamitela*, Linyphiidae, Pauxent, Maryland, USA. GH All photographs by J. Coddington. (From Griswold et al. 1998; reprinted with permission of *Zoological Journal of the Linnean Society*, Blackwell Press.)

For example, if the pre-stress in two otherwise identical webs differs by a factor of two, one web will be two-fold stiffer than the other. Stiff webs absorb insect impact via resistance. Now consider the effects of reduced pre-stress: insect impact can partially be absorbed by web and fiber extension. These effects can be understood by considering the behavior of a thread modeled as a linearly elastic material that is suspended under tension. The axial force in the thread of initial

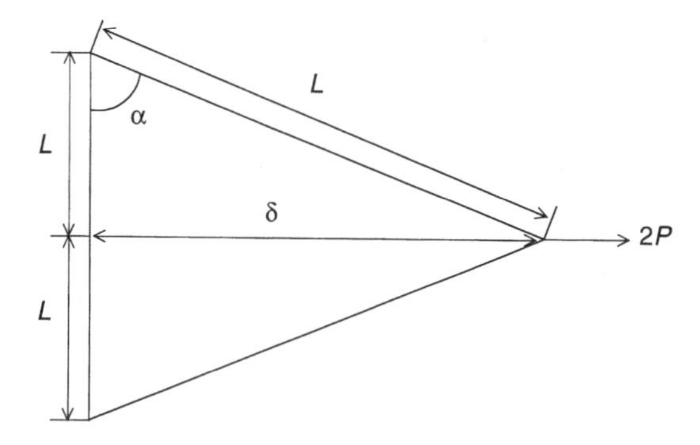

Figure 3.13 The effects of pre-stress on fiber deflection and total stress. Under an applied force of $2P$, the axial force in the thread increases by an amount F_{el}. The total force in the thread is the sum of the pre-stress F_i plus the increase in force due to loading.

length $2L_o$ and pre-stress force F_i will increase by an amount F_{el} if the thread is deflected by a force *2P*. Therefore, the total force in the thread, a function of pre-stress and the applied force, is equal to:

$$F = F_{el} + F_i \tag{1}$$

The elastic modulus (E) of the thread, a measure of material stiffness, is equal to stress/strain:

$$E = (F/A)/((L - L_o)/L) \tag{2}$$

Here, A equals the cross-sectional area of the thread, L_o equals the initial length of the thread, and L equals the fiber's length due to the applied force of *2P*. Rearranging terms, one obtains

$$F = EA(L - L_o)/L \tag{3}$$

For α equal to the angle due to stretch between L_o and L,

$$L = L_o \times sec\ \alpha \tag{4}$$

and

$$\alpha = \arctan\ (\delta/L_o) \tag{5}$$

Therefore, the deflection δ is determined by the equilibrium relationship

$$P = F \times \sin\alpha \tag{6}$$

and the tension in the deflected fiber L is equal to

$$F = P \times \sin\alpha \tag{7}$$

This derivation shows that for a linearly elastic fiber, F is a function of E, the fiber pre-stress, and load position on the fiber. F_{el} and α increase nonlinearly with δ, and the total force F will be larger for a given value of δ when a larger value of F_i is used. Therefore, the lower the fiber pre-stress, the larger the fiber deflection for a fixed load. Flexible webs are more compliant than stiff webs and may be better able to absorb the impact of prey.

Web architectures are often specific to a particular spider genus. Nevertheless, there is no blueprint as to how any single web will be spun and suspended at a particular site. Spiders display considerable behavioral plasticity when choosing among the material properties of potential web supports and potential locations. Spiders that build stiff webs must locate sites where the anchoring structures are stiffer than their silks. Spiders that build compliant webs may adjust the length of framelines in response to varying, ambient air flow as well as the flexibility of the supporting vegetation.

The design properties of webs are also a function of their materials. It could be predicted that webs suspended under high tensions must be built from silks that are inflexible and that deform little when stretched. When an insect is intercepted by this type of web, its impact energy is absorbed, distributed throughout the net, and dissipated into heat. Insect impact is absorbed with less stress on the web's architecture if the silk is elastic or if the web can be displaced. Nevertheless, silks cannot be perfectly elastic. Imagine if an insect were flying into a perfectly elastic web. In the absence of adhesive glues and threads, the insect would be repelled by the web at the same force at which it was intercepted. However, because the web is not perfectly elastic and deforms, there is little resistance against which an insect can leverage itself while trying to escape (Denny 1976).

The breaking energy of a spider silk thread is equal to the product of material toughness times material elasticity. Material toughness is equal to the maximum force a material is able to withstand before breaking. When maximum force per area (F/A; F = Force, A = material cross-sectional area) is plotted as a function of material stress (dL/L_o = stress/strain where L is defined as length), the area under the plot is equal to the maximum amount of energy the material can absorb before breaking (figure 3.14). Therefore, selection can act to

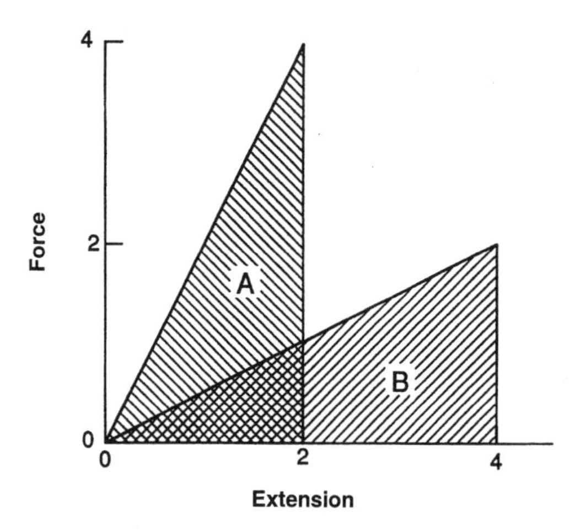

Figure 3.14 Effect of fiber extension on the energy absorption of two materials that deform linearly. When material strength is plotted as a function of material extension, the area under the two curves is equal to the amount of energy the material can absorb. Materials A and B can absorb the same amount of energy but the mechanism of energy absorption differs. Material A absorbs energy primarily through resistance; material B absorbs energy through extension.

affect the breaking energy of silk thread by acting either on fiber strength or on fiber elasticity and stretch. Even if the shapes of the stress/strain curves of two silks differ, the breaking energy of the silks may be equal if the areas plotted under the curves are equal. A weak thread that is extremely elastic may be able to absorb more energy before breaking than a stiff thread that is actually stronger.

The properties of silks, however, differ from the "perfect" or linear elastic materials described in figure 3.14 because the slope of the curve that describes stress/strain varies with degree and speed of extension. Therefore, the stress/strain relationships of silks are nonlinear (figure 3.15). For example, *MA* silks used to produce the frame of spider webs spun by *Araneus sericatus* (these silks are "composites" because they contain both β-pleated sheet, crystalline domains and noncrystalline domains) extend by 30% before breaking. The initial increase of the silk's stress/strain curve is steep, indicating material stiffness of 10^{10} N m^{-2}. Between 2 and 8% extension, the

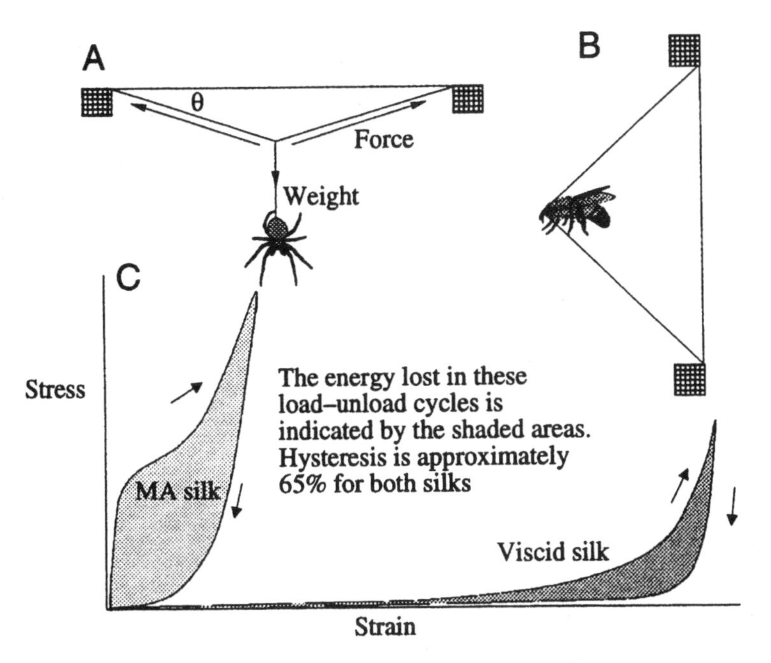

Figure 3.15 Functional loading of *MA* and *Flag* silk. (A) When a static load is applied at right-angles to a silk fiber, extension allows a large deflection angle θ, and the fiber is able to support a large load. (B) When a flying insect impacts a web element, it is stopped by the deflection of a silk that absorbs its kinetic energy ($E_K = 0.5\ MV^2$, where E_K is kinetic energy, *M* is mass, and *V* is velocity) of the insect. (C) The viscoelastic nature of *MA* silk transforms much of insect energy into heat through molecular friction. The energy dissipated is indicated by the area under the shaded loops in the stress–strain curves for load cycle experiments. Hysteresis is the ratio of the energy dissipated to the energy absorbed. (From Gosline et al. 1999 with permission from The Company of Biologists, Ltd.)

Table 3.1 Flag silks are not as strong as *MA* silks but are more extensible. Hence, *Flag* and *MA* silks are equally tough. (From Gosline et al. 1999.)

Material	Stiffness, E_{init} (GPa)	Strength, σ_{max} (GPa)	Extensibility ε_{max}	Toughness (MJ m^{-3})	Hysteresis (%)
Araneus Flag silk	0.003	0.5	2.7	150	65
Araneus MA silk	10	1.1	0.27	160	65
B. mori cocoon	7	0.6	0.18	70	

stress/strain curve plateaus, but after 8% extension the curve again rises steeply, though in this region the silk's material stiffness falls to 4×10^9 N m^{-2} (figure 3.15; table 3.1; Gosline et al. 1984). In comparison, cocoon silk produced by the silk worm *Bombyx mori* extends by only about 20% before breaking. Although its initial stiffness is similar to that of spider frame silk, it breaks at 6×10^8 N m^{-2} (Gosline et al. 1984). Close to 80% of the silk protein spun by *B. mori* is made up of anti-parallel, stacked β-sheets (see chapters 1 and 2). Furthermore, unlike the *MA* silks spun by spiders, cocoon silk contains no residues of either α-helical or coil conformation (Gosline et al. 1984; Simmons et al. 1994). The mechanical properties of cocoon silks may be optimized to maximize stiffness and stretch resulting in a protective shield for developing larvae. Alternatively, the mechanical properties of *MA* silks may be optimized to absorb the kinetic energy of flying prey through fiber extension.

The webs and silks spun by the ancestral Deinopoidea are stiff; their ability to both withstand prey impact and to retain prey is a function of fiber strength

All of the orb webs spun by spiders in Deinopoidea (the superfamily made up of the Uloboridae and Deinopoidea) are horizontal, sticky traps that intercept and retain prey via web and fiber strength. Opell and Bond (Opell and Bond 2001) found that although the breaking energies of catching silks spun by spiders in the Uloboridae and Araneidae are similar, silks spun by uloborids are stronger and stiffer than those spun by the araneids. The lack of fiber flexibility and extension means that all of the insect's impact energy must be absorbed by stresses that develop in the web's support lines. Therefore, the more structural elements that the web contains, such as web radii and framelines, the better it will dissipate energy away from the site of impact (Craig 1987a). Web stiffness and lack of fiber stretch are optimal features of the primitively horizontal orb webs of uloborids.

Uloborids spin their webs close to the ground, in the lower understory or in a crevice of a rotting tree. Hence, these webs are usually in sites that are dimly lit and they are most likely to encounter insects flying either toward or away from an enclosed space. Insects flying through complex vegetation and in dim sites fly more slowly than insects flying in open sites and where light levels are high (Wehner 1981). Relatively low flight speed and body weight result in low kinetic energy and hence low insect impact at the web.

Spiders in only two uloborid genera, *Hyptiotes* and *Miagrammopes*, spin vertically oriented webs. *Hyptiotes* sp. forages in dimly lit forest sites and spins a "triangle" web made up of three sectors of an orb web (Lubin et al. 1978; Opell 1994c). *Miagrammopes* sp. forages at night and spins single capture threads or an irregular network of capture threads that seem to be randomly suspended in both horizontal and vertical orientations. Despite the greater economy of these designs (three sectors of an orb or randomly suspended threads), *Hyptiotes* and *Miagrammopes* actually invest the same amount of material in the web (proportional to spider weight) as do spiders in more ancestral uloborid genera that construct orb webs (figure 3.5; Opell 1996). The greater stickiness and capture area (correlated with spider body size) of derived uloborid webs suggests that selection for insect encounter and retention is a more significant aspect of uloborid foraging strategy than the ability of the web to absorb insect impact. Furthermore, the stickier webs of *Hyptiotes* and *Miagrammopes* have more cribellar fibrils and may reflect more UV light perhaps making them more attractive to insects (see chapter 4).

The diversification of the Araneoidea correlates with a shift in web functional mechanism

Comparison of the integrated function of architecture and silks shows that orb webs and the silks from which they are spun vary in their ability to absorb insect impact energy. On an absolute scale, energy absorption by the webs is a simple correlate of spider size. Therefore, if the material properties of the silks are corrected for spider size differences, there is no difference in silk strength, elasticity or design for webs built by large and small spiders (Craig 1987b). Opell and Bond significantly expanded this view by extending previous analyses to include the ancestral taxa of the Araneoidea (Opell 1997a; Opell 1997b; Opell and Bond 2000; Opell and Bond 2001). These authors showed that the transition from the deinopoid to araneoid webs correlates with a transition in web functional mechanism. Deinopoid webs absorb insect impact primarily through fiber toughness; araneoid webs absorb insect impact via elasticity.

In addition to the araneoids' greater ability to dissipate energy locally through fiber extension and stretch (Opell and Bond 2001), Opell has shown that the evolution of a viscid catching thread correlates with greater material economy as well as superior adhesion of the viscid glues (Opell 1994a; Opell 1994b; Opell 1997a; Opell 1998; Opell 1999b). The materials that make up the araneoid glues include (but are not limited to) inorganic ions, phosphorylated glycoprotein, lipids, and organic, low molecular-weight compounds. The glue is secreted along with the flagelliform thread and encases it when the thread is spun or sheared. When the fiber is relaxed, the glue coalesces into series of regularly spaced droplets. This effect implies that viscid glues are shear thinning fluids. The greater the stress applied to the thread, the better the glue is distributed over it. The greater the impact of the insect, the more effectively the glue spreads over the insect's body and, presumably, the better the insect is retained.

Glue content is extremely variable both within and between species (Higgins et al. 2001), suggesting that glue viscosity could result in an important source of variation on which selection might act. For example, glues produced by spiders that forage in dry environments need to be more resistant to drying than glues produced by spiders that forage in humid forest environments. While the size of the glue droplets correlates with spider size, they may also vary with the light environment in which the spider forages (Craig 1987a; Craig 1988; Opell 1999b).

Silk fibroins produced by derived spiders contain either highly oriented crystalline regions or no crystalline regions at all

Comparative functional analyses correlate the effects of protein structure with silk phenotype. The comparative structural analyses below show how variation in amino acid sequences affects protein configuration. Most araneoid silks contain crystalline and noncrystalline regions. Exceptions are viscid silk and glues which are noncrystalline and, in certain cases, can only be partially crystallized at low temperatures (C.L. Craig and C. Riekel, unpublished). Crystalline regions (regions characterized by long-range order) are generally composed of stacked β-sheets (Fraser and MacRae 1973; Gosline and DeMont 1984; Riekel et al. 1999a). Models proposed for the short-range order of the noncrystalline regions are β-sheet protocrystals of possibly preformed β-sheets (Simmons et al. 1996), β-spirals (Hayashi and Lewis 1998) or coiled, α-helical chains (Dong et al. 1991; Kümmerlen et al. 1996) (see chapters 1 and 2). The different protein domains of long-range and short-range order can be considered as "structural modules." Specific amino acid motifs have been proposed for some of the different structural modules that comprise silks (Hayashi and Lewis 1998; Hayashi and Lewis 2000).

The proportion of the silk protein devoted to each structural module affects the fiber's mechanical behavior (and hence its potential selective value to the organism). The crystalline regions are thought to confer tensile strength to silk polymers. The organization of the noncrystalline regions is probably more variable. One model suggests a matrix of hydrogen-bonded chains that is reinforced by β-crystallites with multifunctional cross-links (Termonia 1994). Such a network would provide fiber strength and elasticity without the added stiffness that derives from β-sheet domains (figure 3.16). For a discussion of the correlation of protein sequence, network topology and mechanical properties see Gosline et al. (1999).

The principal techniques capable of providing atomic scale resolution of the crystalline fraction are transmission electron scattering (diffraction/imaging; TEM) and X-ray diffraction (XRD). Atomic force microscopy (AFM) does not provide the same spatial resolution as the previous techniques, but can be used to image large-scale organization such as fibrillar morphologies (Li et al. 1994; Gould et al. 1999). Comparatively few TEM experiments have been reported on

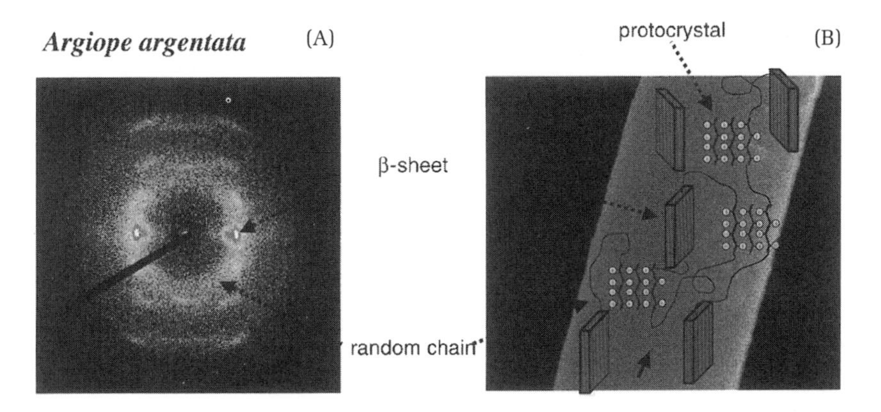

Figure 3.16 Model for the structural organization of *MA* silk. (A) X-ray diffraction (XRD)-pattern for single *MA* fiber produced by *Caerostris sexcuspidata* (C.L. Craig and C. Riekel, unpublished). (B) A model for dragline silks shows β-sheet domains (oriented crystals), protocrystals, and random chains (Simmons et al. 1996). A scanning electron microscopy image of a *N. senigalensis* fiber is shown in the background. The contributions of β-sheet domains (called Bragg peaks) and random chains (identified by the diffuse ring) to the XRD-pattern are highlighted in (A) and idealized in (B). The hypothetical protocrystals would contribute to short-range order close to the Bragg peaks. Models for short-range order and their interplay with long-range order (crystalline fraction) have been proposed (see text). An arrow in this and in the following pictures indicates the direction of the fiber axis.

spider silk (Thiel et al. 1994; Thiel et al. 1997; Barghout et al. 1999; Coddington et al. 2001) which is presumably a result of silk damage due to radiation. In addition, AFM and TEM techniques require elaborate sample preparation techniques such as microtomic slicing. Laboratory XRD techniques can be used *in situ*, but require rather large amount of silks (Warwicker 1960; Work and Morosoff 1982; Parkhe et al. 1997). Therefore, XRD techniques have been used to visualize the dragline silk studies of only a few species (such as *Nephila*) for which sufficient quantities of silk can be obtained by the forced silking technique. Synchrotron radiation (SR) XRD techniques have reduced the number of dragline-fibers required to a few (McNamee et al. 1993). High brilliance X-ray microbeams available at third-generation SR-sources allow the extension of XRD experiments to single silk fibers, threads or composite fiber materials (Riekel et al. 1999a; Riekel 2000). Although currently available SR-beam sizes of a few microns diameter are generally larger than can be attained by electron scattering techniques, the possibility of working on single fibers and in air allows the screening of a range of samples, as will be shown below. Rapid screening provides new opportunities for in-depth experiments on individual silks and experiments with complementary techniques.

XRD-patterns have been gathered for dragline silks (*MA*; major ampullate gland), minor ampullate silks (*MiA*; minor ampullate gland), catching thread silks (*Flag*, flagelliform gland), egg sac silks (*Cy*, cylindrical gland), and silks used to wrap prey and decorate webs (*Ac;* aciniform gland). Additional patterns have been gathered for silks spun by ancestral species, such as the mygalomorph *Antrodiaetus* whose silk glands are acinous in form, and *Euagrus*, whose glands are noted simply as variably sized (Kovoor 1987). These early "proto-silks" are laid into a sheet whose fibers seem to adhere to each other and to prey via an unknown mechanism. They could be described as cotton candy-like: they are easily pulled apart into wisps of extremely sticky fibers. Although the precise mechanism of how they adhere to a surface is unknown, some authors suggest adhesion via electrostatic charge (Peters 1984; Peters 1986).

The evolution of spider spinning behavior (the behavioral process by which proteins are pulled or drawn under tension at secretion; see chapter 1) and fibroin silk correlates with the evolution of araneomorph spiders. Dragline silks, cylindrical gland silks and aciniform gland silks all contain β-sheet crystals to various extents in combination with amorphous material. In analogy to the concept of structural modules (Hayashi and Lewis 1998; Hayashi and Lewis 2000), one can decompose an XRD-pattern into sub-patterns representing long-range and short-range order. This is demonstrated for *MA*-silk spun by the orb-spinner *C. sexcuspidata* in figure 3.17. The entire XRD-pattern for a single fiber is shown in figure 3.17A. The

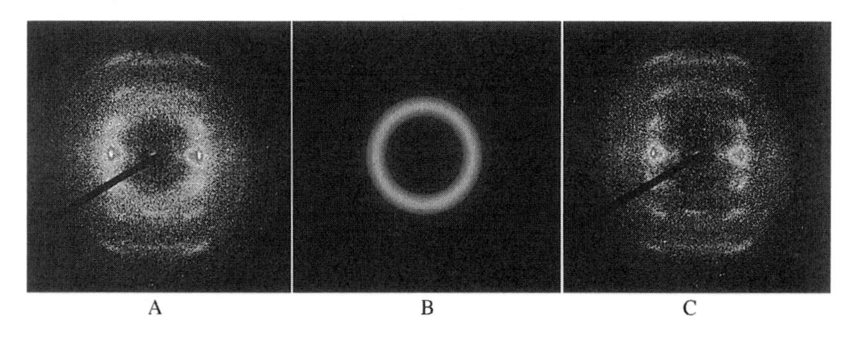

Figure 3.17 XRD-pattern of *MA* silk spun by *C. sexcuspidata* and its separation into structural "modules." (A) Raw XRD-pattern of *MA* thread. (B) Simulated unoriented amorphous phase of silk, illustrating randomly oriented chains. (C) Subtraction of (B) from (A) results in principally an XRD-pattern of the β-sheet phase of the silk. (Same intensity scale for all XRD-patterns.)

amorphous ring in figure 3.17A has been fitted by a two-dimensional Gaussian function and is displayed in figure 3.17B. Finally, figure 3.17C illustrates subtraction of the amorphous ring from the raw pattern and shows, primarily, the crystalline β-sheet phase and some remaining oriented short-range order which can be further separated by fitting routines (Grubb and Jelinski 1997; Riekel et al. 1999a).

The crystalline phase of the majority of the *MA* gland silks corresponds to the β-poly(L-alanine) structure (Marsh et al. 1955a; Arnott et al. 1967) (figure 3.18). For *Nephila senegalensis* orthogonal unit cell parameters of a = 0.93 nm (β-sheet repeat), b = 1.04 nm (interchain repeat), and c = 0.7 nm (intra-chain repeat) were derived (Riekel et al. 1999). Only a few *MA*-silks with the β-poly(L-alanylglycine) structure of *Bombyx mori* (Marsh et al. 1955b) are known until now (see below). A comparison of the two structures and the corresponding molecular organization is shown in figure 3.19.

Comparing XRD-patterns of silks produced by spiders ordered according to their phylogenetic context suggests that silk proteins have evolved from a pattern of:

1. *Ac*-silks produced in acinous glands appear to form β-sheet crystalline domains but have a low crystallinity.
2. *MA*- and *MiA*-silks produced by all araneomorph spiders form highly organized β-sheet, crystalline domains with poly(alanine) or poly(alanylglycine) repeats. The crystallinity is higher than for *Ac*-silks but appears to be variable (figure 3.21). Figure 3.20 shows an *Eriophora fuliginea* XRD-pattern: poly(alanine) and a *Kukulcania hibernalis* XRD-pattern: poly(alanylglycine).

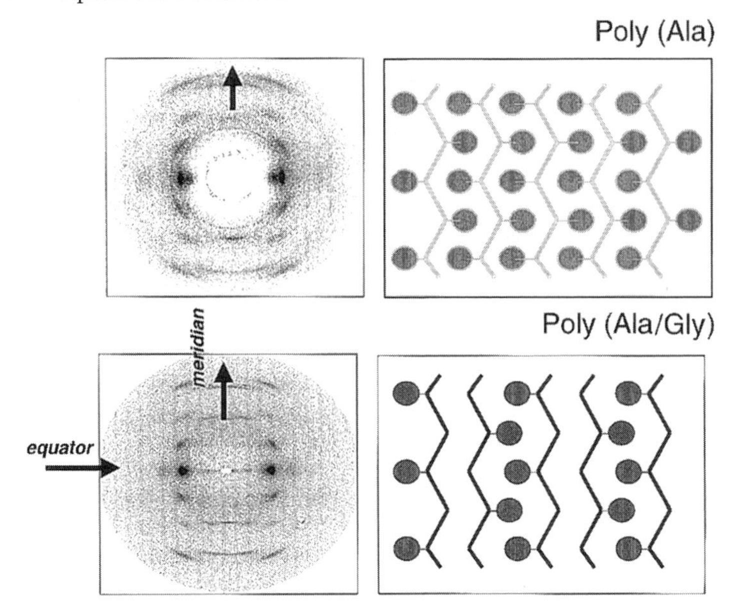

Figure 3.18 Comparison of XRD-patterns. (A) β-poly(L-alanine) XRD-pattern of *Argiope argentata MA* single fiber (Craig et al. 2000) and its corresponding molecular organization. (B) β-poly(L-alanylglycine) XRD-pattern of *Bombyx mori* single fiber and its corresponding molecular organization. In particular, characteristic differences can be seen on the *equator*. The *meridian* is oriented along the fiber direction.

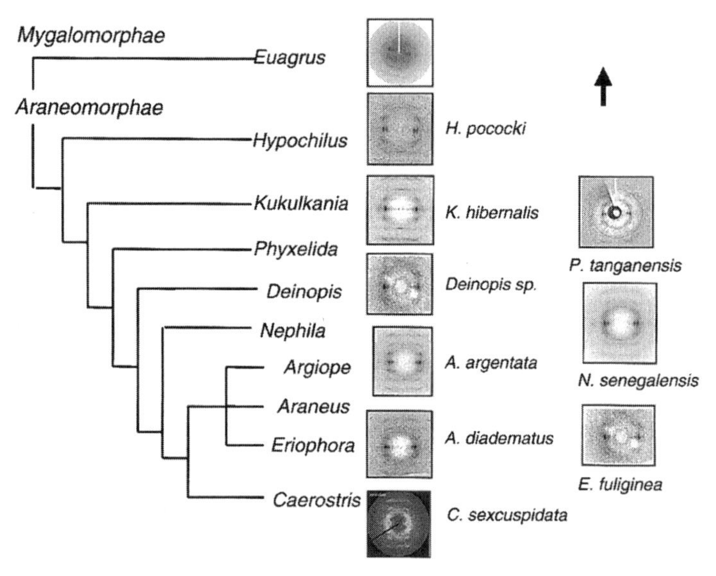

Figure 3.19 XRD-patterns of *MA* silks plotted on an abbreviated phylogeny of the Araneomorphae from Griswold et al. 1998. Tree based on parsimony. Branch support values not available. XRD-patterns show silks spun by *Kukulcania hibernalis* which have a poly(alanylglycine) organization. *MA* silks of other taxa show a poly(A) organization, suggesting that the poly(A) silks are a derived structural organization for *MA* silks.

3. *Flag*-silk produced by the Araneoidea contains little or no crystallinity. In some cases poly(glycine) II type crystallinity is observed (figure 3.21E and figure 3.23C and D).

Although all *MA* silks are characterized by similar XRD-patterns, the structures of the crystalline domains within fibers may differ within and between species (figure 3.20). Thus, dragline silk in *Kukulcania* has a poly(alanylglycine) organization, similar to the silks spun by the silk moth *B. mori*, and in contrast to the poly-(alanine) pattern observed in *Nephila*. These data suggest that dragline silks characterized by pure poly(alanine) crystallites, like those of *Nephila*, evolved more recently and, in particular, correlate with the evolution of complex silk-processing physiology as well as spinning behavior.

According to phylogeny, early silks were less organized at the molecular level than silks spun by the more derived araneomorph spiders. Current XRD data suggest that early silks do not have the range of crystallinities and hence morphological differences available in derived silks. The diameters of early silk fibers are often too small for single-fiber XRD experiments. A more detailed study of local inhomogeneities and orientation of such silks will require X-ray beam sizes extending into the sub-micron range.

Sequence data show that silks, such as those spun by *Euagrus*, contain repeated sequences which include components of the repetitive sequences of amino acids that make up the silks of more derived species. Specifically, these are units of poly(alanine). Even silk spun by the ancestral *Euagrus* shows β-sheet domains (see below; figure 3.21E). In contrast, at least one type of *Flag* silk—that spun by *N. clavipes*—does not seem to contain any poly(alanine) (Gatsey et al. 2001). The poly(A) and poly(A/G) crystalline matrices that impart strength and stiffness to other silk fibroins are largely lost

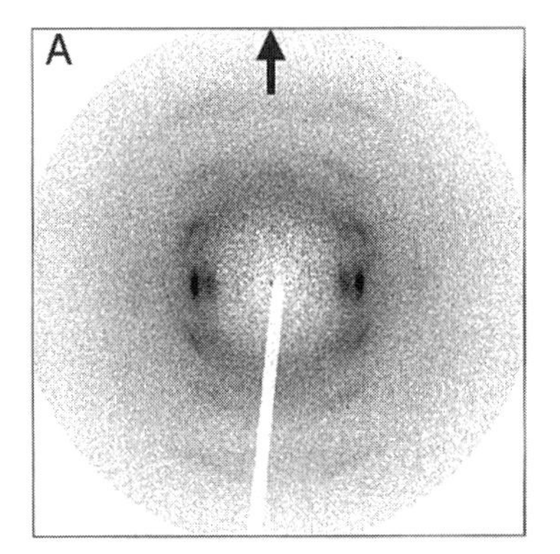

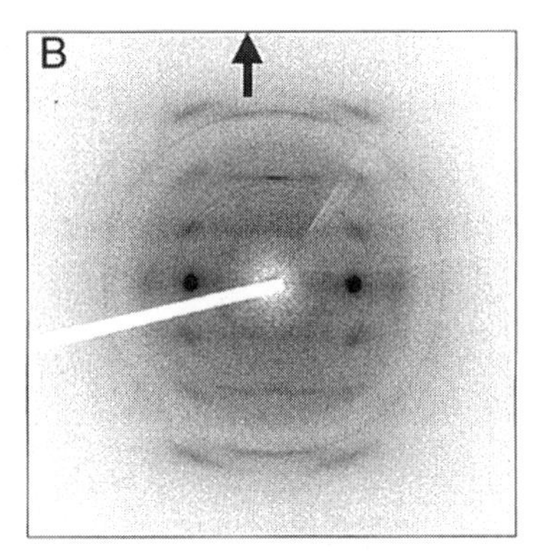

Figure 3.20 XRD-patterns of (A) *Eriophora fuliginea* silk with poly(alanine) structure; and (B) *Kukulcania hibernalis* silk with poly(alanylglycine) structure (compare also with figure 3.18).

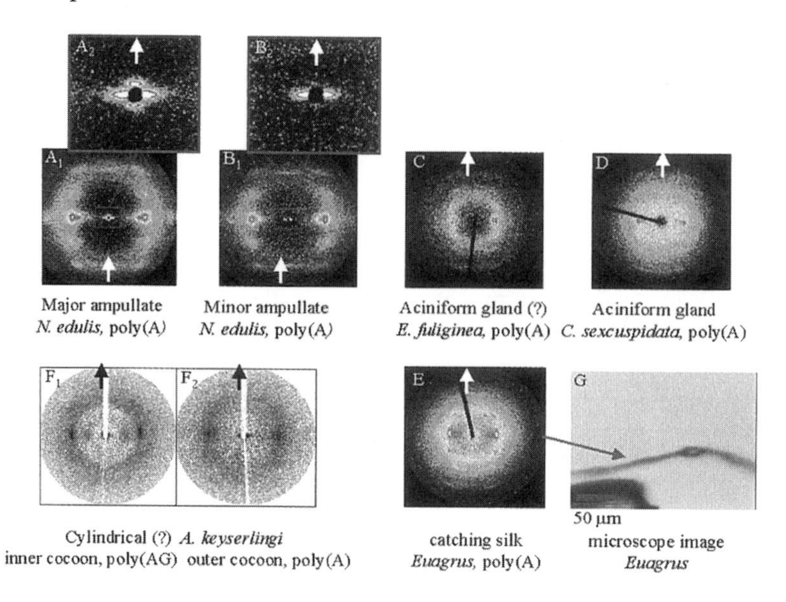

Figure 3.21 Selected XRD-patterns from silks produced by various glands. *Major ampullate* silk: (A_1) *N. edulis;* (A_2) zoom showing SAXS-range with an ≈8 nm meridional peak (Riekel and Vollrath 2001); (B_1) Minor ampullate silk: *N. edulis;* (B_1) zoom showing SAXS-range (Riekel and Vollrath 2001). Aciniform silk: (C) *E. fuliginea;* (D) *C. sexcuspidata, Euagrus;* (F) *Argiope keyserlingi.* (C.L. Craig and C. Riekel, unpublished). The *Euagrus* XRD-pattern (E) was obtained outside the fibrous glue droplet; see microscope image (G).

and replaced by a new type of organization that imparts strength but also results in fiber flexibility and stretch.

Major ampullate silk

Dragline silk spun by the golden orb-weaver, *Nephila clavipes*, has a composite structure containing crystalline domains equaling roughly 30% of the dry silk volume embedded in a noncrystalline matrix that makes up the remaining 70% volume (Gosline et al. 1984). XRD data suggest the existence of two amorphous fractions of different degrees of orientation (Grubb and Jelinski 1997; Riekel and Vollrath 2001). The random chains generate a ring-like XRD-pattern, which centers radially at the position of the strongest equatorial β-sheet reflection. The observation of a meridional reflection (spacing 8 nm) by X-ray small-angle scattering (SAXS) (figure 3.21A_2) implies that crystalline and amorphous domains form a fibrillar morphology with an approximately 8 nm axial repeat (Yang et al. 1997; Riekel and Vollrath 2001). The equatorial streak has been associated with a model of (idealized) cylindrical fibrils (Yang et al. 1997; Riekel and Vollrath 2001). Evidence for microfibrils has also been obtained by

AFM (Li et al. 1994). This suggests a three-phase model with fibrils composed of crystalline and amorphous domains embedded in a matrix of more randomly oriented hydrogen-bonded chains. As indicated above, the amorphous fraction might have a more complicated short-range order.

Minor ampullate silk

Silk drawn from the minor ampullate gland (*MiA*) of *Nephila edulis* is thought to be more crystalline than silk drawn from the major ampullate gland (Guerette et al. 1996) (figure 3.21B_1). Specifically, *Nephila MiA* silks contain relatively more (GA) couplets than do the *MiA* silks of *Araneus diadematus* (Gatsey et al. 2001). XRD on dry *MiA* silk from *Eriophora fuliginea* shows very little short-range order, suggesting an expected high crystallinity (Riekel et al. 2001). In-situ XRD data from *N. edulis* silk during forced silking, however, shows a significant short-range order although its crystallinity appears to be higher in *N. edulis MA*-silk (Riekel and Vollrath 2001). Setting aside real differences in crystallinity between *MiA*-silks from different species, a heterogeneous composition (Stubbs et al. 1992) could complicate interpretation of the patterns. A better distinction of *MA* and *MiA*-silks appears to be possible based on SAXS-data (figure 3.21A_2 and B_2). Both types of silk show the equatorial streak due to fibrillar morphology (Grubb and Jelinski 1997; Riekel and Vollrath 2001). The meridional reflection is, however, only observed for *MA*-silk which can be explained by a higher volume fraction of crystalline poly(alanine) domains in the fibrils.

Aciniform silk

Spiders use aciniform silks to wrap prey and to decorate their webs (see chapters 5 and 6). XRD data show that aciniform silk is less crystalline than spider dragline silk, although its diffraction pattern shows that it contains β-sheet domains with poly(alanine) sequence (figure 3.21C,D). Aciniform silks, however, have notably less alanine (ratio Ala/Gly = 1) than do ampullate gland silks (Ala/Gly is ~0.5). Aciniform gland silks also have a high serine content, and this may account for the apparently high volume of noncrystalline material. Simmons et al. (1996), proposed a 2-D folding pattern for dragline silk in which the protein chain folds back on itself at Gly-Ser connections. The high serine content of aciniform silk introduces more frequent folding, thereby reducing the degree of long-range order in the thread. This might explain the larger amorphous fraction in aciniform silk than in dragline silk, although a more complicated morphology cannot be excluded. The short-range order appears, however, to vary between species. Thus, aciniform silk of *E.*

fuliginea has short-range order centered on the strongest reflections (figure 3.21C) while short-range order in *C. sexcuspidata* is more homogeneously distributed (figure 3.21D).The model of aciniform proteins—that is, reduced crystalline and high volume of non-crystalline material—suggests that this protein, like the proteins produced in the aciniform glands of the ancestral spiders, are functionally protein glue. It may be that spiders can regulate the fraction of the protein devoted to alanine or alanine/glycine couplets. Some investigators believe that all of the silk glands of spiders evolved from an acinous like ancestor (J. Kovoor, pers. commun.). It is therefore of interest that the ancestral spider *Euagrus* shows a similar β-sheet diffraction pattern with short-range order as for *C. sexcuspidata* (figure 3.22E).

Figure 3.22A shows an optical image of *E. fuliginea* wrapping silk. The morphology of the flexible fiber wound around the threads allows assigning it to an *MiA*-fiber. XRD with a 5 μm beam of the two nearly horizontal threads shows a significant short-range order.

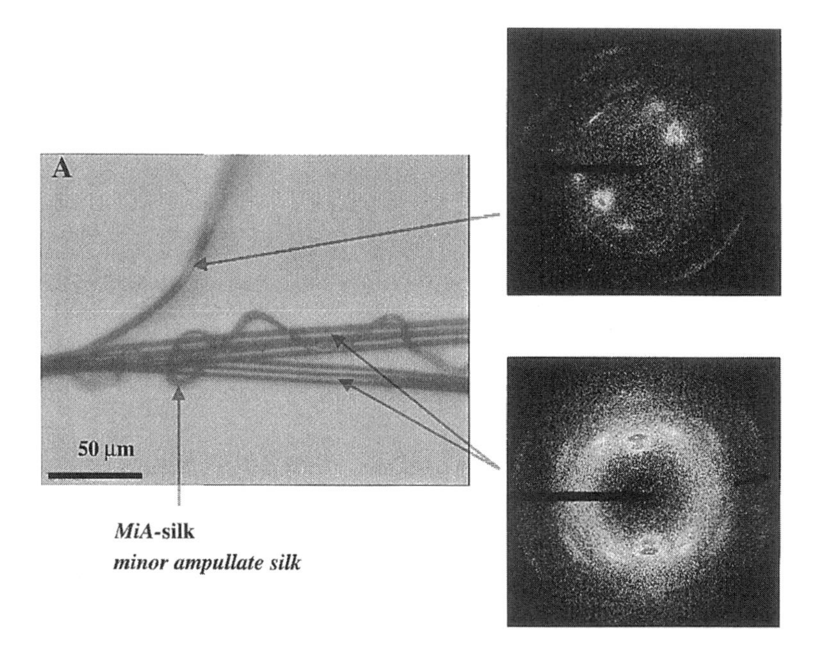

Figure 3.22 Silk classified as *E. fuliginea* "wrapping silk" is composed of different types of silks. (A) Optical image: morphological differences allow distinguishing MiA-silk (coiled around threads) from several threads (the upward pointing fiber is a thread viewed side-on). (B) XRD with a 5 μm beam allows distinguishing differences in crystallinity between the individual threads. The two XRD-patterns have to be horizontally flipped so that the meridien direction corresponds to the macroscopic fiber axis. The *Ac*-XRD-pattern shown corresponds to the lower thread.

These threads are assigned therefore tentatively to *aciniform* silk as they have the lowest crystallinity. The upward-pointing thread (side view) shows very little short-range order and is tentatively assigned to *MA*-silk. Such assignments will become firmer as the database of investigated silk increases. Experiments with X-ray beams extending into the micron- and submicron-range (Müller et al. 2000; Riekel 2000) should also allow a better investigation of possible composite morphologies (Stubbs et al. 1992).

Cylindrical gland silk

The cylindrical glands are present in most of the araneomorph spiders. These glands first appeared in *Telema tenella* (Telemidae, eyeless cave spiders) and *Filistata insidiatrix* (Filistatidae) (J. Kovoor, pers. commun.). They are also present in *Uroctea durandi* and appear as numerous simple tubes. Cylindrical gland silks seem to have a higher alanine/glycine ratio than *MiA* silks. They are rich in serine and glutamic acid, which may have a structural influence.

XRD experiments on an *Argiope keyserlingi* cocoon show structurally different silks (figure 3.21F_1 and F_2) (C.L. Craig and C. Riekel, unpublished). The inner part of the cocoon is made of *Bombyx mori*-type poly(alanylglycine) silk, while the outer part is composed of *MA*-type poly(alanine) silk (see figure 3.18). It is not known, however, whether the cylindrical glands produce both types of silks. Nevertheless, figure 3.21 shows that the spider is capable of spinning different types of silks in order to obtain a specific functional material. The optimization of stiffness in cocoon silks has already been mentioned above. Another protecting feature might be that poly(alanylglycine) silk is less hydrophilic than poly(alanine) silk and could thus protect the eggs from water.

Flagelliform silk, pseudoflagelliform silk and viscid silk glue

Pseudoflagelliform glands are not present in Hypochilidae, Thaididae, Eresidae, Oecobiidae, Amaurobiidae, and Filistatidae (J. Kovoor, pers. commun.). Only the derived cribellate spiders have pseudoflagelliform glands, and only the araneoid spiders have flagelliform glands. Furthermore, both the flagelliform (or pseudoflagelliform) probably evolved in correlation with aggregate (or cribellar, paracribellar) glands as these two gland types form a functional unit (J. Kovoor, pers. commun.).

The fibers of pseudoflagelliform gland silks contain both amorphous and crystalline material. Brushed silk drawn from the pseudoflagelliform gland of *Uloborus glomosus* is characterized by very low crystallinity. The fibers are too small for single-fiber XRD. The powder ring on top of an amorphous halo (figure 3.23A) corresponds in position to the strongest β-sheet reflection of *Nephila* (lattice spacing ~0.44 nm). This material is similar to silk sampled from the catching silk of the more ancestral non-orb-spinning cribellates, the *Oecobiidae*. The radial thread silk from *U. glomosus* shows a much higher orientation (figure 3.23B). This silk appears not to belong to poly(alanine)-type silks as it does not show a characteristic β-sheet reflection (d ≈ 0.44 nm). A detailed structural analysis has not yet been performed.

Flagelliform gland silk is the most recently evolved type of silk and *Flag* silk spun by *N. clavipes* shows basically no crystallinity. *Flag* silk spun by *E. fuliginea* shows at 100 K a partially crystalline XRD-pattern (figure 3.23C) (C.L. Craig and C. Riekel, unpublished). A similar XRD-pattern (although of lower crystallinity) is observed for *C. sexcuspidata* (figure 3.23D). The crystalline fraction of both patterns resembles the poly(glycine) II structure (Crick and Rich 1955). The helical structure determined for poly(glycine) II (Crick and Rich 1955) is of interest with respect to proposed helical models for short-range order in spider silk (see page 70, and Chapters 1 and 2).

XRD-patterns from viscid droplets that coat the *Flag* silks and act as fibrous glue show short-range order which peaks at the position of the short-range order peak of dragline silk (figure 3.23E, F, and G).

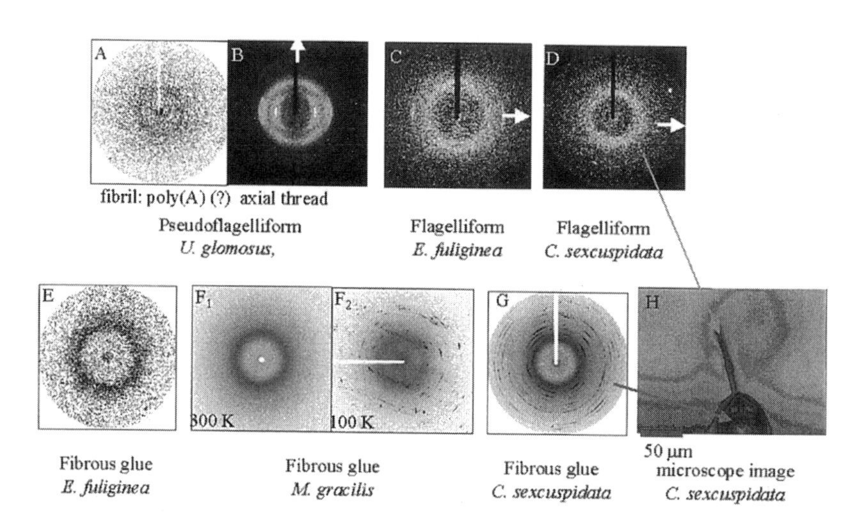

Figure 3.23 Selected XRD-patterns for pseudoflagelliform, flagelliform silk and protein glues. (A,B) Pseudoflagelliform silk spun by *U. glomosus*; (C) flagelliform silk spun by *E. fuliginea*; (D) flagelliform silk spun by *C. sexcuspidata*; (E) fibrous glue produced by *E. fuliginea*; (F_1,F_2) fibrous glue produced *M. gracilis*; (G) fibrous glue produced by *C. sexcuspidata*; (H) microscope image of droplet *C. sexcuspidata*. (From C.L. Craig and C. Riekel, unpublished.)

The examples shown for *E. fuliginea, Micrathena gracilis* and *C. sexcuspidata* show differences in short-range order, which suggest differences in protein composition. Thus, *E. fuliginea* fibrous glue has a weak peak at smaller angles (d = 0.98 nm), which is not observed for *M. gracilis*. A crystalline fraction is observed for *C. sexcuspidata* fibrous glue at room temperature. *M. gracilis* fibrous glue was found to have a partially crystallized fraction at 100 K (figure $3.23F_2$), while data obtained from a different sample at 300 K show only short-range order (figure $3.23F_1$). The diffraction spots appearing at low temperatures suggest the presence of crystallites, which are much larger than the 3- to 6-nm crystalline domains determined for dragline silks. A similar crystalline fraction has also been observed for a viscid droplet in *Cyclosa conica* silk (C.L. Craig and C. Riekel, unpublished). It is probable that these crystallites are innate to the fibrous glue, as the strongest peaks coincide with the short-range order peak and the *M. gracilis* crystallites form at low temperature. The observation of larger crystallites in viscid silk droplets suggests the possibility of investigating low temperature phase transformations in other silks, notably *MA* silks. It is of interest to note in this context that low-temperature electron microscopic examination of *N. clavipes MA* silk has suggested the presence of crystallites of 70–100 nm diameter (Thiel et al. 1994).

Flag silks are as strong as *MA* silks, but more elastic and more flexible (table 3.1). Their flexibility, and hence compliance, is a key factor correlated with their ability to absorb prey impact (Opell and Bond 2001). The lack of β-pleated crystallites may be significant to the silk's high elasticity.

Despite the advantages of araneoid webs and silks, the cribellate spiders have persisted through evolution

At least twenty-two families of spiders spin dry or cribellate capture threads. This type of silk is produced by spiders in almost all of the major lineages of spiders (Coddington and Levi 1991). Spiders that produce cribellate silks primarily spin webs that are irregular. Only the Deinopoidea spin orbs from cribellate silks. Comparison between deinopoid and araneoid orb webs suggests that araneoid silks and webs are far superior with respect to functional capabilities. As discussed above, araneoid webs have greater flexibility and elasticity than deinopoid orbs. *Flag* catching threads intercept larger and faster flying prey than deinopoid catching threads. Furthermore, araneoid webs are simpler to build and metabolically cheaper to produce than webs spun by the deinopoids (Opell 1997b; Bond and Opell 1998; Opell 1998; Opell and Bond 2001). The 32-fold increase in species number that correlates with the araneoid divergence from the deinopoids has resulted in significant habitat expansion, increase in body size, diversification of reproductive strategies, and life histories. Taken together, all of these factors suggest that the araneoids should out-compete the deinopoids and other cribellate spiders, thereby driving them to extinction. Nevertheless, the cribellates have persisted. Is there something unique about cribellate silks, webs, or the types of habitats that these spiders use, that differentiates them from the araneoids? This something may be related to the behavior of prey in the presence of webs and is the subject of the next three chapters.

Summary

The *MA* silks correlate with the emergence of the Araneomorphae, or 93% of all known spider species (figure 3.24). Although *MA* silks are not the only factor that correlates with the emergence of the Araneomorphae, they would have had a significant effect on the ecological opportunities available to spiders. The *MA* silks contain highly organized, β-pleated structures. Furthermore, *MA* silks are used without the addition of other materials. The evolution of fibroin

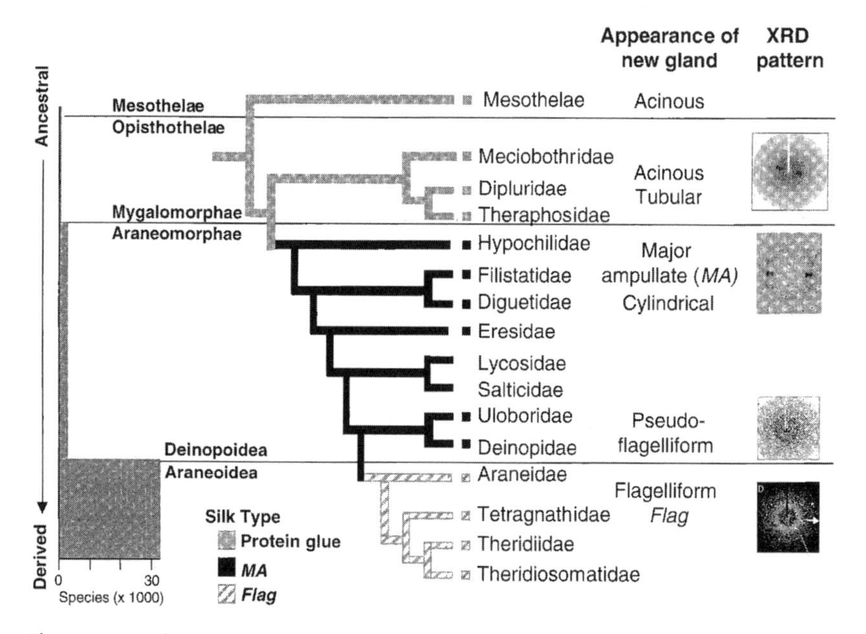

Figure 3.24 At least three major speciation events correlate with the evolution of new types of silk glands and new fibers. (Data from Bond and Opell 2000.) Composite phylogeny from Goloboff 1993; Griswold, et al. 1998; Scharff and Coddington 1997. No branch support values available.

silks in spiders is analogous in importance to the evolution of flight in insects and affords spiders many of the same ecological opportunities. For example, because single *MA* threads are weight-bearing, spiders are able to use them for dispersal as well as to support vertically suspended, capture nets. Furthermore, even though *MA* silks may differ in amino acid content, a broad, cross-taxa comparison of their crystalline organization suggests that most of this variation is attributed to the noncrystalline regions of the protein.

The *Flag* silks correlate with the emergence of the Araneoidea, the web-spinning spiders. This was the second major burst of spider speciation, giving rise to a 32-fold increase in species number (Craig et al. 1994; Bond and Opell 1998). *Flag* silks have a molecular organization, sequence organization, and X-ray diffraction pattern unique among all the other silks and fibrous proteins studied to date. The X-ray patterns show either only short-range order or short-range order together with crystalline domains resembling the helical organization of poly(glycine) II. *Flag* silk spun by *A. diadematus* is three orders of magnitude more flexible than *MA* silk (Gosline et al. 1999). The flexibility and elasticity of *MA* and *Flag* silks make the webs spun by the araneoids unique.

4

Insect Spatial Vision Is a Potential Selective Factor on the Evolution of Silk Achromatic Properties and Web Architecture

with M. Lehrer

When judging the possible role of natural selection in the evolution of the properties of spider silks, one must think not only of the silks, but also of the visual capacities of the spiders' prey, the insects. Put simply, why do some insects fly into spider webs whereas others are clearly able to avoid them? To investigate this question, we consider the insects' responses to both spatial and chromatic parameters of webs and silks. In the present chapter, we will focus on the insects' spatial vision and the spatial parameters of webs. The insects' color vision and the chromatic parameters of silks will be dealt with in chapter 5.

Insect spatial resolution is a function of the anatomy and the optics of the eye

All insects possess so-called compound eyes; so called because they are made up of a complex array of tiny units called ommatidia (facets; figure 4.1). Depending on the insect species, the compound eye contains between several dozens and 10,000 ommatidia, each between 10 and 40 μm in diameter. Each ommatidium has its own lens, its light-absorbing structures, and its neural connections to the

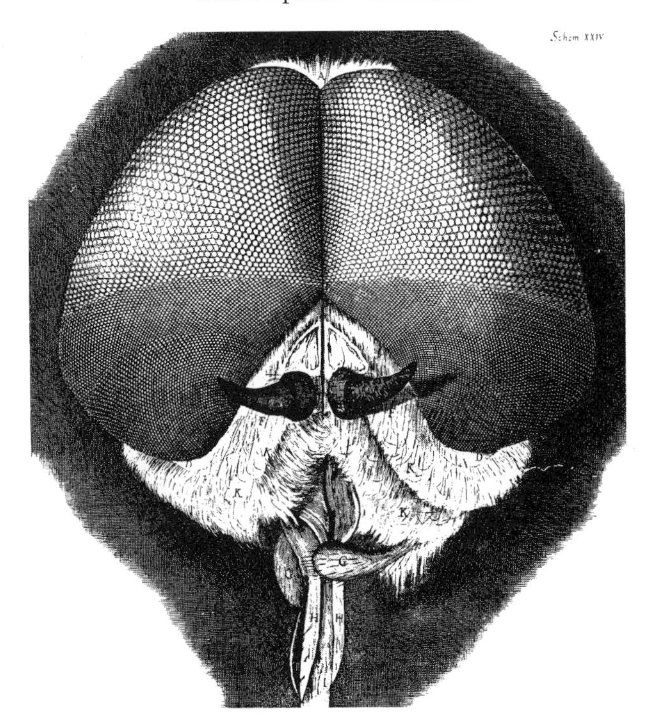

Figure 4.1 The insect eye is made up of a complex array of sampling stations or ommatidia (From *Micrographia*, Hooke 1665; original figure kindly contributed by R. Wehner). The variation in diameter and curvature of the facets in the eye of a tabanid fly result in regions of different visual acuity and contrast sensitivity.

brain. The ommatidial grid transmits to the brain a mosaic-like image, the grain of which depends on the density of the ommatidial array, that is, on the number of "sampling stations" per area of the eye surface.

A single ommatidium does not form an individual image. The only information it receives is the number of light quanta, that is, the intensity of the light it absorbs. Two objects, one dark and one bright, can be resolved only if they project on two different (neighboring) ommatidia. Thus, the smaller the distance between neighboring ommatidia, the finer the grain of the image. This distance is measured as the angle, $\Delta\phi$, subtended between the visual axes of two neighboring ommatidia. Depending on the insect species, $\Delta\phi$ varies between 1° and 5° (see Table 1 in Wehner 1981). For comparison, in the human retina sampling density is approximately 100-fold greater. Thus, the spatial resolution power of the insect compound eye is much poorer than that of a vertebrate single-lens eye.

Small values of $\Delta\phi$ usually go hand in hand with small facet diameters and are thus equivalent to a high sampling density of the eye. There is a limit, however, to how small the diameter of an ommatidium can be. A small ommatidium absorbs less light and therefore it perceives a lower light intensity than a larger ommatidium. As a result, the difference in brightness (that is, the contrast) between two neighboring ommatidia becomes too small to be detected. This implies that spatial acuity depends not only on the sampling density of the photoreceptor array, defined anatomically by the interommatidial angle $\Delta\phi$ mentioned above, but also on the contrast sensitivity of the eye, measured as the acceptance angle $\Delta\rho$ of a single ommatidium. The acceptance angle is an optical parameter that determines the size of the visual field of the ommatidium and therefore the amount of light absorbed in it. (For more details, see Snyder 1979; Stavenga 1979; Wehner 1981; Land 1989; Land 1997.) During the course of evolution, a compromise had to be made between fine-grain imaging at the cost of contrast sensitivity, and a high contrast sensitivity at the cost of detail richness.

The differences in eye design found among different insect species very likely represent adaptations to various types of habitat. Mainly, optimization of both spatial acuity and contrast sensitivity been the factors determining facet size (Goldsmith 1990). In general, insects that forage in dim light have a larger facet size and higher contrast sensitivity, but they view a coarser image than insects that forage in bright light. Some insects that forage in more complex or multiple light environments have eyes that are divided into regions of multiple facet sizes; this allows the animal to use its eyes efficiently even under changing local environmental conditions (figure 4.1). For a recent review, see Warrant (2001).

Contrast resolution is the prerequisite of object detection

The difference in (subjective) brightness between two visual stimuli is perceived as intensity contrast when the two are viewed next to each other. The intensity contrast contained in two adjacent stimuli, also termed intensity modulation (m), is defined as the ratio between the amplitude of the modulation and the mean intensity of the two stimuli. This ratio can also be expressed as:

$$|m| = |I_1 - I_2|/(I_1 + I_2) \qquad (1)$$

where I_1 is the intensity of one of the two stimuli, and I_2 is that of the other. (For measurement and calculation of intensity values, see Srinivasan and Lehrer 1988.) The contrast m can take on any value between zero and 1, or, expressed in percentage, between 0% (when $I_1 = I_2$), and 100% (when $I_2 = 0$). The highest possible intensity contrast is contained in the combination black and white. An object viewed against an equally bright background (m close to zero) is invisible, unless the two differ in color (see chapter 5).

Usually, the visual scene contains more than just one contrasting object. The density at which contrasting areas occur within the image is expressed by the so-called contrast frequency of the pattern. Contrast frequency is determined by the number of contours contained in a unit area of the image. A contour is the edge between a dark and a bright area, or between two areas that differ in color. The larger the number of contours, and thus the smaller the distance between neighboring contours, the higher is the contrast frequency of the pattern. For example, a white square containing five black bars has a higher contrast frequency than a square of the same size containing only three bars. In repetitive patterns such as linear or radial gratings, spatial frequency is measured as the number of cycles (or periods) of the pattern per unit length, or per visual angle subtended at the eye. In behavioral experiments, bees were shown to discriminate very well between patterns that differ in contrast frequency (for a review, see Lehrer 1997a; for a review of the older literature, see Wehner 1981).

Insects possess a high temporal resolution capacity

In insects, the visual system and the motor system have been intimately linked throughout their evolution. This is because insect eyes are fixed so that movement of the head or the body cannot be compensated by counter-motion of the eyes. Thus, every flight maneuver causes motion of the image at the eye. During motion, the contrast frequency of the image viewed is defined as the number of contours or of cycles passing at the eye per unit of time. When an insect is flying fast in a patterned environment, the contrast frequency of the image at the eye may reach very high values. The insect's survival crucially depends on its capacity to resolve moving images.

A detailed study on swarming and mating in the minute fly, *Anarte pritchardi* (Cecydomyiidae), conducted by Okubo et al. (1977) illustrated the astounding capacity of this fly to maneuver and orient, even at extremely high flight speeds. The Cecydomyiidae are thought to be the family of flies that make up the most abundant and important food source for small orb-spinning spiders in the tropics (for example, small araneoids such as *Leucauge*, *Epilineutes*) that live in the tropical forest understory (Craig 1989b). Okubo et al. induced the flies to swarm, filming them with a high-speed camera. The flight pattern and speed were then analyzed in three dimensions using both the position of the insect and the position of the insect's shadow. When a female entered the swarm, the males' response was instantaneous: in an attempt to mate, a male *A. pritchardi* accelerated as fast as 90 cm s^{-2} (4.9×10^3 body lengths per second) to catch up with the female. Correlated with the increased speed was an increase in oscillation about the fly's forward path. The speed and agility with which the cecydomiids are able to maneuver in a three-dimensional environment and still perceive visual objects makes one wonder how spiders are able to capture any of these insects at all. It seems that the webs that intercept insects either produce no contrast against their background and are therefore invisible to the insect, or they are considered to be harmless if the insect does see them.

More recent results provide further evidence for insects' excellent motion resolution capacity. Honeybees were found to perceive the rotation of black-and-white radial gratings up to a contrast frequency of 200 Hz (Srinivasan and Lehrer 1984), a behavioral response that is as fast as is the photoreceptors' response to flickering light stimuli (Autrum and Stöcker 1950). In humans, the cut-off frequency of motion resolution lies at about 50 Hz. The insects' ability to resolve images of high temporal frequency suggests that a spider web, if it is visible to the insect, will also be visible during flight as the insect approaches the web.

The important role that the insect's high temporal resolution might play in the present context is evident from results of experiments videotaping *Drosophila* (Diptera) as they approached webs of *Mangora pia* (Araneoidea: Araneidae) and *Epilineutes globosus* (Araneoidea: Theriodomatidae). Webs spun by *M. pia* are characterized by a high fiber density (about ten fibers per centimeter), whereas webs spun by *E. globosus* are characterized by low fiber density (four threads per centimeter; figure 4.2). Assuming that eye anatomy alone determines spatial resolution, we can calculate the maximum distance d at which the insect can resolve the web by using:

$$\tan(\alpha/2) = h/2d \qquad (2)$$

where α is the just resolvable visual angle, and h is the distance between two silk fibers. If we take $\Delta\phi$ in *Drosophila* to be 2.5°, then the limit of resolution $2\Delta\phi$ (see above) would be 5°, and the calculation renders $d = 11.4$ mm for *M. pia* ($h = 1$ mm) and $d =$ 28.6 mm for a web spun by *E. globosus* ($h = 2.5$ mm). In the experiment, however, no interceptions were observed. The fly turned away from the web of *M. pia* at a mean distance of 70 mm. In other words, the flying insect could make out the web at a greater distance then it could have been resolved and or predicted under stationary viewing conditions. Thus, image motion seems to enhance object visibility and insects can avoid webs even if the web details are not identified.

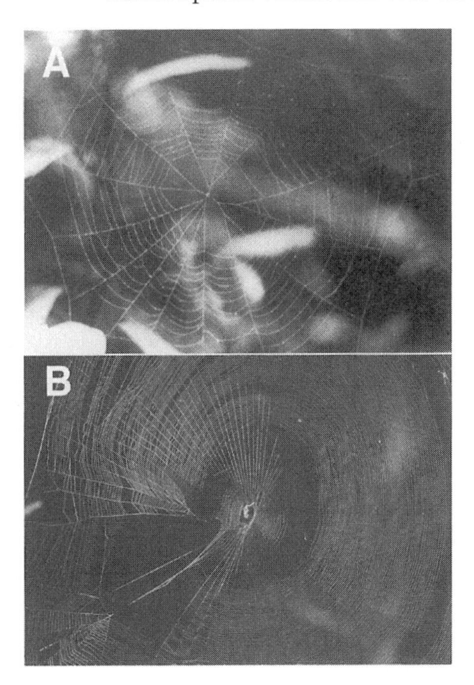

Figure 4.2 High- and low-fiber-density webs. (A) Webs spun by *Epilineutes globosus* (Theridiosomatidae) are characterized by low-fiber-density; the distance between two spiral turns is about 2.4 mm. (B) Webs spun by *Mangora pia* (Araneidae) are characterized as high-fiber-density webs; the distance between two spiral turns is about 1 mm. *D. melanogaster* must fly much closer to a web spun by *M. pia* to resolve its pattern (11.4 mm) than a web spun by *Epilineutes globosus* (28.6 mm). Nevertheless, the fly avoided the high-fiber density web from a mean distance of 70 mm (see text).

Evaluation of flight paths videotaped in this experiment (figure 4.3) suggests that the insect improves its resolution power by actively producing image motion. Shown are two flight paths of *D. melanogaster* approaching webs spun by *E. globosus*. In one diagram, the insect flies on a relatively smooth path (figure 4.3B). In the second diagram, the insect approaches with an erratic flight path and actually touches the web before accelerating away (figure 4.3). It may be that the fly's increased flight oscillation frequency as it approaches the web traces a visual path that allows the insect to better resolve the web pattern and thus avoid capture. Similarly, the increased distance from which *Culex quinquefasciatus* (Diptera; not illustrated) was able to avoid the web (about twice the distance found in the fruit flies) might have been due to the high contrast frequency resulting from unsteady flight. In the honeybee, active acquisition of motion cues was demonstrated several times (see Lehrer and Srinivasan 1994; Lehrer 1996).

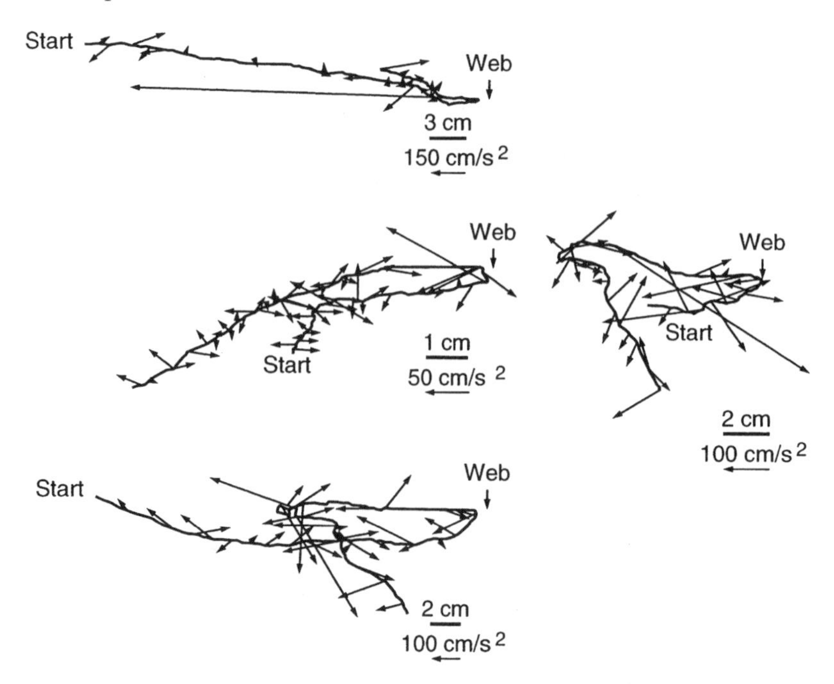

Figure 4.3 Acceleration vectors plotted around the approach paths of *D. melanogaster* as they fly towards a lower-fiber density web spun by *E. globosus.* Top left: The small size of the displacement vectors around the fly's flight path indicates smooth flight. When *D. melanogaster* reaches the web area, however, it rapidly accelerates in the opposite direction before touching it. Middle, Bottom left: The large size of displacement vectors around the flies' paths correlate with unsteady flight. Here two different *D. melanogaster* approach the web and before touching it, accelerate in the opposite direction to avoid encounter. Right: A fourth fly approaches the web, touches it, and accelerates in the opposite direction to avoid capture. (Redrawn from Craig 1986.)

Motion parallax cues provide the insect with depth information

As pointed out above, image motion perceived during flight has implications for the insects' power of temporal resolution. However, it does more than that—namely, it provides the insect with depth information.

Most insects possess only small overlapping binocular visual fields and are therefore incapable of binocular three-dimensional vision (Burkhardt et al. 1973; Frantsevich and Pischka 1977; Seidl and Kaiser 1981). Furthermore, three-dimensional vision based on triangulation is impossible due to the two eyes being situated close together (Burkhardt et al. 1973; Collett and Harkness 1982). Instead, insects extract information on the third dimension (that is, on the distances of objects in the environment) by exploiting the apparent speed of image motion. During free locomotion, stationary objects move faster at the eye when they are near than when they are farther away. Honeybees were shown to be very adept at discriminating among different speeds of image motion (Lehrer et al. 1988). Indeed, the most important role that image motion plays in insect orientation involves the acquisition of depth cues based on the speed at which contours move at the eye (for reviews, see Srinivasan et al. 1989; Lehrer 1996). Detection of the edge between two contrasting areas is very relevant in the context of spatial vision because it constitutes the first step towards detecting an object against its background (Srinivasan et al. 1989; Campan and Lehrer 2002).

In the context of web avoidance, the role of depth estimation based on motion cues was evident in experiments in which the distance between web and background was varied (Craig 1990). In those experiments, more insects were intercepted when the web was close to the background than when it was placed at a greater distance. Indeed, the amount of relative motion between web and background decreases when the distance between the two is decreased.

Insect vision and flight maneuverability function as potential selective forces on silk and web properties

Several investigators have assessed orb web design and function on the basis of prey size alone (Pointing 1966; Robinson and Robinson 1970; Robinson and Robinson 1973; Nyffeler and Bentz 1978; Uetz et al. 1978; Riechert and Luzak 1982; Wise and Barata 1983; Eberhard 1986). This approach, called the collision model of prey capture (Craig et al. 1985), predicts that the spaces between turns of an orb's viscid thread are about equal to the size of the spiders' most common prey. Hence, insects are sieved by webs from the air streams that pass through them (Kajak 1965; Dabrowska and Luczak 1968; Kiritini et al. 1972; Nyffeler and Bentz 1978; Riechert and Luzak 1982). While the collision model may adequately describe prey capture in some cases, it does not in others. Using artificial webs, Nentwig (1983) showed that the mean size of the intercepted insects was independent of the grid size when the prey were small.

The problem with the collision model lies in the assumption that all orb webs function identically and that the insects approaching them are passive; that is, they are unable to control their flight behavior and to detect webs. This assumption is contradicted by the observations mentioned above (Okubo 1973; Okubo et al. 1977), and by several field observations of insects approaching and avoiding webs and, in some cases, even flying through them (Lubin 1973; Chacon and Eberhard 1980; Rypstra 1982; Craig 1986).

The insect's "effective size" depends not only on the length of its legs, but in addition, on its flight pattern. Complex flight patterns that take the insect on a trajectory that is longer than would be necessary, often changing direction or flying on arcs and spirals, increase the insect's effective size. Furthermore, although such trajectories, by producing image motion, enhance object detection (see above), they result in a decreased amount of forward flight. The consequence is an increase in the time that an insect spends within the web area and where interception is probable (Craig et al. 1985).

Building on the studies of Okubo, Craig (1986) videotaped *Drosophila melanogaster* (Cyclorrhapha: Drosphilidae), *Culex quinquefasciatus* (Nematocerae: Culicidae), and *Trichoprosopon digitatum* (Nematocerae: Culicidae) in free flight as they approached spider webs. These insects were chosen because they differ in their habitats and morphology. The fruit fly possesses a small leg-to-body-length ratio, which makes it relatively compact. The two mosquito species, on the other hand, have a larger leg-to-body-length ratio and often fly with their legs extended. The long legs may increase the effective size of the insect, thus enhancing interception.

Each of the three insect species chosen for the study was filmed in free flight without airflow, and with airflow of 1.4 m s^{-1} against the insect's flight direction. Insects of all three species maintained the same average ground speed in both conditions. In the presence of airflow, however, the two mosquito species had a higher air speed, in agreement with the increase in oscillation frequency. *C. quinquefasciatus* maintained the direction of its flight path by bursts of acceleration in alternating directions. In *D. melanogaster*, on the other hand, air speed as well as ground speed were kept constant in either condition, and only small directional adjustments took place during flight (figure 4.4). Thus, at least in that study, the effective size of the insect was not correlated with morphological size or with airspeed, suggesting that the insect's actual size does not influence its flight pattern, nor does it predict the probability of interception.

Distorted and oscillating webs may enhance insect interception

The insects' specialized visual capacities and their flight performance described above suggest that their potential to avoid interception by spider webs is great. Not surprisingly, web-weaving spiders seem to have evolved means to reduce the visibility of their webs. Web visibility depends in part on the light conditions in which the spider forages. Nevertheless, in any site, the ability to produce translucent silks would seem to be a logical step toward the evolution of a cryptic web. However, another strategy might

D. melanogaster

Air speed = 17.5 cm/s

5 cm

250 cm/s 2

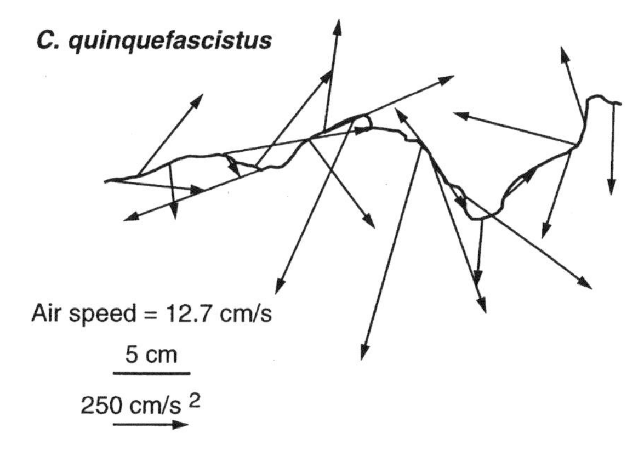

T. digitatum

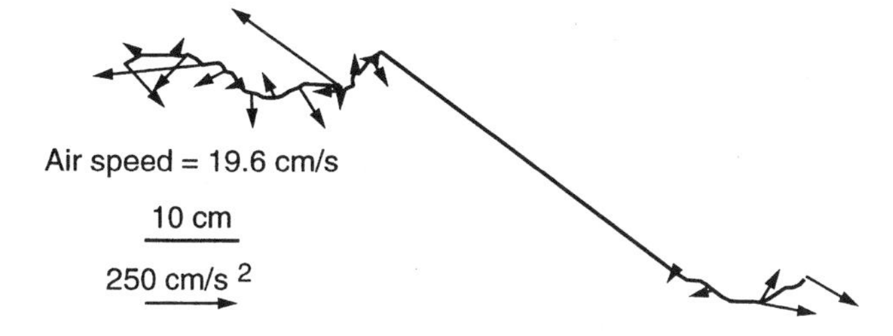

Figure 4.4 Comparison of flight paths of *C. quinquefasciatus, T. digitatum,* and *D. melanogaster* when flying at about the same speed. During forward flight, *C. quinquefasciatus* maintains its flight path by large bursts of acceleration in alternating directions. As a result, its effective size is much larger than its true size. In contrast, *T. digitatum* and *D. melanogaster* maintain constant air and ground speeds while making only small directional adjustments during flight. In this case, insect "effective size" is closer to its true size. (Redrawn from Craig 1986.)

also be effective. Due to the small size of insects relative to the web size, even if an insect can resolve the web, at short range only a small portion of the web will be visible. Design specializations that distort the plane of the web might render the layout of the web complicated and unpredictable in the eyes of an inspecting insect.

An approaching insect attempting to fly through a hole or around the web could be intercepted. Indeed, several spider species make distorted webs that might serve to confuse insects. There are two ways, dynamic and static, in which spatial distortions of the web plane can be achieved.

Dynamic distortions

Dynamic distortion of the web plane is achieved by the specific architecture of the web and by the airflow surrounding it. These two factors are the basis of the "encounter model" of prey capture (Craig et al. 1985). For example, webs suspended on flexible supports oscillate even in low airflow (figure 4.5). The oscillatory patterns are erratic, enhancing the probability of insect capture over a volume of space, and not just within one wing or leg length from the web surface. Furthermore, the success of insects attempting to fly through webs will be reduced, because erratic vibrations of single threads or sections of the web may cause them to contact the flying insect. Often, the oscillations of the web are superimposed on the insect's active efforts to acquire motion parallax information by performing erratic flight maneuvers, thus increasing the insect's effective size, and thus the probability of interception even further.

At stationary webs built by small spiders, the relationship between the insect's effective size and the spacing between fibers of the spiral turn may determine the probability of safe passage (which is

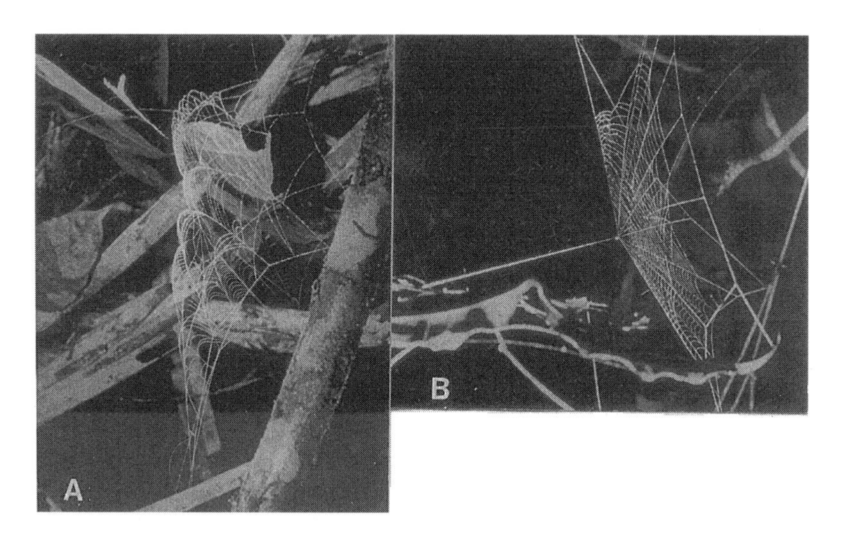

Figure 4.5 Dynamic and statically distorted orb webs. (A) Oscillating web spun by *E. globosus*. (B) Static web displacement of web spun by *Epeirotypus* sp. (From Craig 1986.) Reprinted with permission from Elsevier Science.

consistent with the collision model). However, at an oscillating web, the effective size of the web is as relevant as the effective size of the insect, their interactions making safe passage considerably less likely (encounter model). In general, only relatively small insects try to fly through webs, suggesting that this type of orb web, spun by small spiders, may have evolved capturing very small prey. For example, *Leucauge globosa* (Araneidae; adult female, 2.7 mg) spins multiple webs per day, which fluctuate over a total distance of about 80 mm. In contrast, the larger spider, *Micarthena schreibersi* (Araneidae; adult female weighs 146 mg) spins one web per day, which displaces over a maximum distance of 0.5 cm (Craig et al. 1985).

Static distortions

Although only small spiders spin oscillating webs, not all webs built by small spiders oscillate. In addition, some large spiders seem to modify their webs in ways that prevent oscillation. For example, spiders in the orb-spinning genus *Nephila* often surround their webs with a network of supporting lines that limit the web's movement. Other large orb weavers either spin their webs in protected sites or spin short support lines that allow the spiders to fit webs into open spaces in the vegetation. When vegetation in a potential site obstructs webs spun by *Argiope argentata*, the spider may actually spin silk fibers that draw branches and leaves away from the web surface. The supports on such webs are comparatively short and, as a result, the web surface is often distorted when fit into a site. It may be that an approaching insect, or one hovering in front of the web to inspect it, will only see a portion of the web within the limited range that it can resolve. Regions lying outside the resolvable plane may not be visible.

Spiders of the family Theridiosomatidae take active web distortion a step further. *Theridiosoma* sp. and *Epeirotypus* sp. both spin small orb webs and suspend a fiber from the center of the web to the vegetation behind it. The spider sits in the center of the web, facing away from the web surface, and then pulls the bridging fiber taut. As a result, the web surface is distorted into a cone that is 1 to 4 cm deep. Due to their limited range of resolution, insects viewing the web may not be able to perceive the web center. Given the readiness of *D. melanogaster* to fly through holes in highly visible webs, it may be that insects take the visual hole, due to the extension of the web into a cone, to be an open pathway. Additional confusion might result when the spider releases the central fiber that distorts the web and causes it to fling toward the insect. Static distortion of the web plane may enhance prey capture significantly in sites where airflow is low, such as the forest understory where *Theridiosoma* and *Epeirotypus* are found (Craig 1986).

The translucent properties of frame (*MA*) and spiral (*Flag*) silk minimize contrast between webs and their background

The silks that make up the radii of araneoid spider's orb-webs refract, reflect, and absorb light. Refracted light is slowed and bent as it passes through the fiber. From most angles, web radii are difficult to see; because silks are refractive materials, if light strikes a web fiber at a 90° angle, it will be highly visible. Vertically suspended webs viewed in the early morning or late afternoon when the sun is low will appear brilliant because the light reflected from the web plane is maximized (figure 4.6). Insect activity, at least at the webs spun by forest spiders such as *M. pia*, and nonforest spiders such as *A. argentata*, is highest between dawn and 10 a.m., precisely when the web is most likely to be visible. Webs suspended in diffuse light, where the light has no strong directional component, are difficult to see (Craig 1988). In the case of *Epilineutes globosus*, its prey are most active during the light rain that proceeds a tropical downpour, hence when light in the understory is diffuse (C.L. Craig, pers. observ.).

Perhaps due to their lack of crystalline subdomains (see chapters 2 and 3), the reflectance properties of the silk that makes up the araneoid web spiral (Flag silk) is even less visible than that of the web's support fibers. However, unlike *MA* silk, spiral silk is coated with spherical droplets of a glycoprotein or glue that functions to retain

Figure 4.6 Orb web spun by *Micrathena* sp. in early morning sun. The viscid droplets that coat the web's spiral fiber disperse and scatter light, making the web visible to prey. The visibility of the web changes constantly due to the movement of overstory vegetation and the changing position of the sun. Whereas illuminated portions of the web may be well visible and draw prey to spider foraging sites, shadows obscure the web outline, making the pattern of the web difficult to see. (From Craig 1989; reprinted with permission from Blackwell Publishing.)

intercepted prey (see chapter 3). The viscid droplets coating Flag silk are the primary feature that makes orb webs visible. The density of individual droplets and their size depend on how the silk is pulled from the spider's spinneret (Walker 1987). Unlike silks, the viscid droplets scatter light broadly, making the web spiral more reflective and visible than the web supports, even though the silk itself may be invisible. For example, when measured in the laboratory, the spiral fiber spun by *A. argentata*, a spider that forages in bright light, has a strong directional component and appears bright only at relatively acute angles. In natural conditions, *A. argentata* webs are usually suspended in open sunlight and are almost always difficult to see. In contrast, the viscid fiber produced by *M. pia* (that forages in the forest understory where light is diffuse) scatters light much more broadly when examined in the laboratory. In the field, however, the web is suspended in dim sites where light rarely strikes it. When it is struck by a sunfleck (C.L. Craig, unpublished), the web appears brilliant and is clearly visible (analogous to *Micrathena* in figure 4.6). These differences in brightness could be a by-product of selection for protein glues that function in specific habitats. The reflectance properties of the glues may also correlate with the light environments in which different spider species are found. It is conceivable that the droplets, originally meant as glue to retain prey, have (through their particular reflectance properties) taken on a secondary function, namely to attract insects to the web. This idea was tested by videotaping freely flying stingless bees, *Tetragonisca angustula*, as they approached, avoided, or were intercepted at spider webs under constant light conditions.

Orb-webs spun by *N. clavipes* were collected on plastic rings, misted with water, and allowed to dry. The result was the breakdown of droplets on the web spiral. Hence, it was possible to test insect response to webs that were identical in design and silk spectral properties but which differed in the total amount of light they reflected due to the presence or absence of the droplets. The videotapes revealed that in the absence of a web, but in the presence of an open ring, 52% of the bees that approached the ring flew through it. When the ring was replaced by an untreated web, however, only 3.4% of the bees that approached it flew into it. In contrast, when the open ring was replaced by a water-washed web, 11% of the approaching bees flew into it. An analysis of variance for categorical data and linear contrasts shows that the bees responded significantly differently to the control rings and the rings with webs ($P = 0.0001$). Although the number of insects approaching treated webs ($n = 202$) was almost identical to the number of bees approaching the control rings ($n = 205$), significantly more insects ($n = 296$) approached the untreated webs or webs in which viscid droplets made them more visible than the control rings ($n = 205$). The data suggest that, although natural webs would actually be easier to avoid due to the better visibility of the viscid droplets, the same droplets—which are shiny and sparkling due to their shape—may attract the insects' attention and draw them to the web area.

Insects' response to webs is independent of ambient light conditions

If the insects' visual system (or their perception of webs) represents an important selective factor on the optical properties of silks, then insect response to spider orb webs should be generalizable; that is, it should remain constant regardless of the particular diurnal and nocturnal light conditions under which the insect encounters the web. A series of experiments was designed to test this idea by comparing insect response to webs that were made to appear brighter than usual in dim, fluctuating, and bright ambient light (Craig 1988).

During the rainy season, the species of orb-web weaving spiders found on Barro Colorado Island (BCI; a field site of the Smithsonian Tropical Research Institute, Panama) are abundant and their light environments are diverse. The response of diurnal insects to webs spun by *M. pia* (Chamberline and Ivie), found in dense understory vegetation and representing a dim light habitat; *Micrathena schreibersi*, found in open understory and forest gaps and representing a fluctuating light habitat; and *Argiope argentata*, found in nonforest sites and representing a bright light habitat, were measured (Craig 1988). In addition, insect response to webs of two nocturnal species were also measured: *Eustala* sp. 1, found in nocturnal, nonforest sites (variable light, depending on moon and clouds); and *Verrucosa* sp., a nocturnal spider found in nocturnal forest sites (that are constantly dark).

The reflectance properties of half of each web were enhanced by dusting them with a mixture of cornstarch (to enhance reflectance) and confectioner's sugar (water absorption by sugar maintains web stickiness) (figure 4.7). The other half of the web was left untreated. The relative rates of insect interception at different web sites during the same time period were monitored by sequentially mapping and later comparing the accumulation of damage. This made it possible to determine whether webs that are sighted are avoided.

The comparison of web damage that resulted from insect interception showed that in all of the diurnal light conditions, and even at nocturnal sites under a full moon, insects were more successful in avoiding the more highly reflective web halves than the untreated control halves (Wilcoxon matched-pairs signed-ranks test, $P < 0.001$ for all diurnal sites; $P < 0.005$ for *Eustala* foraging under a full moon). In forest sites, where some nocturnal webs are never exposed to moonlight, and in nonforest sites in the absence of a moon, silk

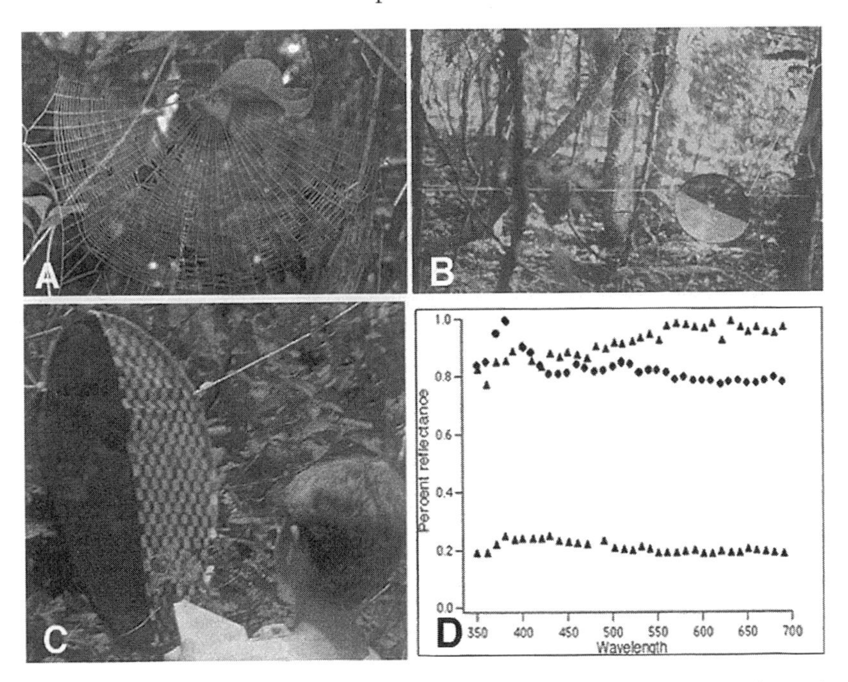

Figure 4.7 Experimental approaches to affect contrast between webs and their backgrounds and determine their effects on rates of prey interception. (A) One-half of an orb web spun by *Mangora pia* is dusted with cornstarch and sugar to enhance silk reflectance; the remaining, untreated half serves as control to determine the effects of silk reflectance on web visibility to prey. (B) A solid, half-black-and-white background is suspended behind a web spun by *Micrathena schreibersi* to determine the effects of high and low background contrast on web visibility to prey. (C) A half-checked, half-solid black background is suspended behind a web spun by *Argiope argentata* to determine the effects of background pattern on web visibility. (D) Reflectance spectra for starch and sugar, white material and black material. The reflectance spectra of the black paint used to make the patterned backgrounds was identical to that of the black material and is not shown. In all cases, the damage patterns on the webs were mapped over at least a four-hour period to determine the relative rate of insect interception at the experimental or control web halves. These experiments show that insects will avoid webs if they see them. (From Craig 1988; Craig and Freeman 1990. Reprinted with permission from Blackwell Publishing.)

reflectance had no effect on insect-web encounter. These results suggest that web visibility is an important factor affecting the foraging performance of at least some nocturnal and diurnally foraging spiders. They show, in addition, that the insects used no odor cues to detect the web.

The studies outlined above also suggest that the viscid droplets that coat the web spiral (see previous section) are the most important factor determining web visibility. Unlike the silks, the droplets reflect and scatter all of the light that strikes them. However, in addition to their effect on visibility (that can be either advantageous or disadvantageous to the spider), the primary function of the viscid droplets that coat the web spiral is to retain prey. Therefore, web visibility must also be correlated with stickiness and hence be subject to an alternative and perhaps conflicting set of selection pressures. An across-species and across-habitat comparison of silks is one approach to determining broad patterns of properties that characterize viscid silks.

To this end, we collected viscid silks from webs spun by twenty-three species of orb-spinning spiders in sixteen genera and, through a series of pairwise tests, compared their brightness (Craig and Freeman 1991; figure 4.8). The sampled silks were from webs spun by spiders whose body sizes varied over an order of magnitude, spiders that forage at night or during the day, and spiders that forage in the forest understory, in forest gaps, and in open, nonforest sites. In addition, we also compared silks spun by conspecifics of different sizes, as well as spider species of different sizes within the same genus (Craig and Freeman 1991). The silks were ranked from 1 to 23 for silk brightness and sorted according to droplet size, spider size, and light habitat (table 4.1).

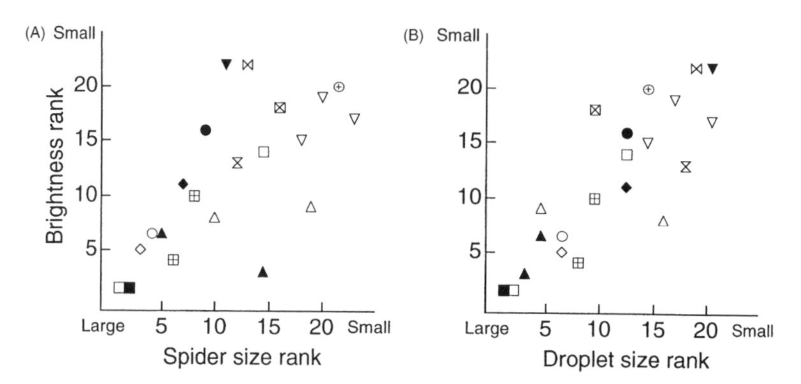

Figure 4.8 (A) Plot of silk brightness rank as a function of spider size rank for spiders in 16 different genera. (B) Plot of brightness rank as a function of droplet size rank. Symbols: (vertical cross in box) *Araneus;* (open circle) *Micrathena;* (open, up-pointing triangle) *Witica;* (open, down-pointing triangle) *Leucauge;* (solid, up-pointing triangle) *Metazygia;* (horizontal butterfly) *Cyclosa;* (open square) *Nephila clavipes;* (solid square) *Eriophora;* (corner-to-corner cross in box) *Alpaida;* (solid diamond) *Verrucosa;* (solid circle) *Eustala;* (cross in circle) *Pronous;* (vertical butterfly) *Mangora;* (solid, down-pointing triangle) *Tetragnatha.* The viscid droplets that coat the catching fiber of webs spun by orb weavers in the families Araneidae and Tetragnathidae are closely correlated with size ranks of large spiders (ranks 1–8) ($r = 0.69$, $P < 0.05$), but vary among small spiders (size ranks 9–23) ($r = 0.22$, $P > 0.05$). With the exception of silks produced by *Metazygia* sp. 2, the data show that the relationship between the size of the viscid droplets and spider body size is more closely correlated among nocturnal spiders ($r = 1.0$, $P < 0.001$) than among diurnal foragers ($r = 0.66$, $P < 0.005$) (see tables 4.1 and 4.2). (Modified from Craig and Freeman 1990.)

Table 4.1 Spider identification and data used for silk and spider rankings, in addition to habitat type and foraging times. (From Craig and Freeman 1990)

	Silk brightness rank	Spider size (mm)	Droplet size (mm)	Habitat type	Foraging period
Eriophora	1.5	23.1	7.3×10^{-2} SD = 6×10^{-3} ($n = 2$)	F	N
Nephila clavipes	1.5	29.6	4.5×10^{-2} SD = 2.7×10^{-3} ($n = 4$)	F	D
Metazygia sp. 1	3	6.7	3.0×10^{-2} SD = 1.0×10^{-3} ($n = 6$)	O	N
Araneus sp. 1	4	11.0	2.06×10^{-2} SD = 3×10^{-4} ($n = 5$)	F	D
Argiope argentata	5	20.2	2.7×10^{-2} SD = 4×10^{-3} ($n = 5$)	O	D
Metazygia sp. 2	6.5	12.4	2.9×10^{-2} SD = 3×10^{-4} ($n = 7$)	O	N
Micrathena sexspinosa sp. 1	6.5	15.7	2.7×10^{-2} SD = 3×10^{-3} ($n = 3$)	F	D
Witica sp. 1	8	8.8	9.3×10^{-3} SD = 4×10^{-4} ($n = 4$)	F	D
Witica sp. 2	9	5.8	2.9×10^{-2} SD = 1.3×10^{-3} ($n = 3$)	F	D
Araneus sp. 2	10	9.5	1.7×10^{-2} SD = 4×10^{-4} ($n = 7$)	O	D
Verrucosa	11	9.7	1.5×10^{-2} SD = 1×10^{-3} ($n = 7$)	F	N
Cyclosa sp.	12	4.5	3.7×10^{-3} SD = 3×10^{-4} ($n = 8$)	F	D
Mangora pia	13	7.0	8.4×10^{-3} SD = 2×10^{-4} ($n = 6$)	F	D
Micrathena sp.	14	6.7	1.5×10^{-2} SD = 1.3×10^{-3} ($n = 5$)	F	D
Leucauge sp. 1	15	6.0	1.1×10^{-2} SD = 4×10^{-4} ($n = 3$)	F	D
Eustala	16	9.2	1.5×10^{-2} SD = 1×10^{-3} ($n = 7$)	F	N
Leucauge globosa	17	3.1	5.3×10^{-3} SD = 5×10^{-4} ($n = 6$)	F	D
Alpaida	18	6.4	1.7×10^{-2} SD = 4×10^{-4} ($n = 2$)	F	D
Leucauge sp. 2	19	5.0	8.9×10^{-3} SD = 9×10^{-4} ($n = 6$)	F	D
Pronous	20	4.5	1.1×10^{-2} SD = 1.3×10^{-3} ($n = 5$)	F	D
Tetragnatha	22	7.6	5.3×10^{-3} SD = 4×10^{-4} ($n = 11$)	O	N
Acacesia	22	6.1	2.7×10^{-3} SD = 1×10^{-4} ($n = 6$)	O	N
Cyclosa caroli	22	6.9	8×10^{-3} SD = 4×10^{-4} ($n = 4$)	F	D

F, forest; O, nonforest; D, diurnal; N, nocturnal.

Table 4.2 Nonparametric regression technique, Tukey's resistant line, is calculated for each case below. The Spearman rank correlation coefficient and significance value are also calculated for the data in each case (this correlation coefficient does *not* refer to the fit of Tukey's resistant line, but the relationship between points in each data set).

Droplet size rank = 4.26 + 0.52 (spider size rank); $r_s = 0.62$, $P < 0.001$
Silk brightness rank = 2.28 + 0.43 (droplet size rank); $r_s = 0.81$, $P \ll 0.001$
Silk brightness for 8 largest species
Silk brightness = 3.03 + 0.4 (spider size rank); $r_s = 0.82$, $P \ll 0.01$
Silk brightness for 15 smallest species
Silk brightness = 12.7 + 0.1 (spider size rank); $r_s = 0.03$, $P \ll 0.05$
Silk brightness for diurnal species
Silk brightness rank = 3.04 + 1.3 (spider size rank); $r_s = 0.66$, $P < 0.005$
Silk brightness for nocturnal species
Silk brightness rank = 1 (spider size rank); $r_s = 1.0$, $P < 0.001$

The Spearman Rank Correlation Coefficient test shows that silk brightness is closely correlated with droplet size ($r_S = 0.81$, $P << 0.001$; figure 4.8A; table 4.2) and that spiders which forage at night produce silks that are significantly brighter than those produced by spiders of the same size but which forage during the day (figure 4.8). Furthermore, large spiders produce droplets, the sizes of which are closely correlated with spider size ($r_S = 0.82$, $P < 0.05$), although somewhat unexpectedly in small spiders there is no such correlation ($r_S = 0.03$, $P > 0.05$; figure 4.8B). These relationships were constant within some genera, but varied within others. For example, silk brightness correlated with the body size of spiders in the genera *Araneus*, *Micrathena*, and *Witica*. On the other hand, among the genera *Leucauge*, *Metazygia*, and *Cyclosa*, silks spun by smaller-sized spiders were brighter than silks spun by larger-sized spiders. When ranked according to natural light conditions, a relationship between silk brightness and spider size is clear only in those sites where the ambient light is predictable. Specifically, silk brightness appears to be directly correlated with the body size of spiders that are found in dim, nocturnal sites. In general, nocturnal spiders produce silks that are brighter than those of the diurnal spiders. This result might indicate that when light levels are low, the amount of light reflected from the silk is so small that web visibility does not constitute a factor affecting selection of web properties. Therefore, by producing relatively dense, sticky glues, the spider may be more successful in retaining intercepted prey than would be the case at a web suspended in a brighter site.

Web visibility is determined by specific web-background combinations in specific ambient light conditions

A web is visible to insects only if it produces contrast against its background. In an experiment which was complementary to that described above, Craig (1990) manipulated the contrast between the web and its background once more, but this time by varying the brightness of the background, rather than that of the web (see figure 4.7B). Artificial web backgrounds were suspended behind natural webs in each of the three light habitats used in the previous experiment. The backgrounds, made to web size, consisted of a Plexiglas hoop covered with bridal veil. Half of each background disc was covered with solid white veil, and half with solid black veil. The difference between the two halves consists in the sign of contrast that a web will produce against each of them: regardless of the web used, it is predicted to appear darker than the white background, but brighter than the black one. As in the previous experiments, we compared insect responses to the web halves by mapping the locations on the web where insects were intercepted. We found that there was no difference in insect response to either half of the web, regardless of the sign of the contrast and or the ambient light conditions.

Insects also detect the contrast of objects against textured backgrounds, mainly based on relative motion (see for example Srinivasan et al. 1991; Campan and Lehrer 2002). Insects were found to be attracted to areas of high contrast frequency (Hertz 1930; Free 1970; Anderson 1997) and thus they might be attracted to the complex textures of vegetation. Some investigators found a positive correlation between the density of webs spun by orb-weavers and shrub-foliage diversity (Der Ver Sholes and Rawlins 1979; Hatley and MacMahon 1980; Rypstra 1983), as well as species-specific differences in the distance of webs from their background (Enders 1974; Enders 1977; Lubin 1978; Uetz et al. 1978; Olive 1980; Brown 1981; Wise 1981; Spiller 1984; Craig 1988). Thus, a larger number of insects might be captured at webs suspended in front of specific types of vegetation, and some vegetation backgrounds might obscure web outline better than do others.

Even at sites with comparable light conditions, different backgrounds, such as patterned and unpatterned ones, are likely to affect web visibility to insects. Spiders indeed seem to exploit the properties of the background when selecting the sites for their webs. For example, *Argiope trifasciata* and *Araneus trifolium* are two species that coexist in old fields under constantly bright light conditions. The relatively large and visible webs spun by *A. trifasciata* are usually found close to the ground (Robinson and Robinson 1970; Enders 1977; Olive 1980; McReynolds and Polis 1987), spun against vegetation that may obscure the web outline. On the other hand, the small and more loosely spun webs produced by *A. trifolium* are suspended close to the tips of old field shrubs (Olive 1980), against more homogeneous or distant backgrounds. In both cases, the appropriate choice of the background ensures that the web remains invisible against its particular background.

We tested for the effect of background pattern once again by suspending a background disc behind the web, but this time modifying it so that half of the disc was covered with black veiling and the other half with black-and-white checkered veiling. In these experiments, insect response varied with the ambient light. In sites where light levels were high (that is, nonforest sites), web halves that were suspended in front of a patterned background were more difficult for insects to avoid than webs suspended in front of a solid background (see figure 4.7C). In dim fluctuating light, background pattern had no effect on web visibility.

The results of these experiments motivated a series of laboratory experiments in which the background, and the light intensity, could be more precisely controlled. Specifically, the aim was to determine: (i) which background patterns better obscured web outlines; (ii) how the background-to-web distance affected insect perception; and (iii) how light levels interacted with background pattern and background distance to web to affect the ability of insects to perceive the web (Craig 1990).

The laboratory experiments were conducted in a flight chamber and in light with a spectral distribution similar to that of natural sunlight (Craig 1990). *D. melanogaster* flies were introduced into a darkened end of the chamber and video-taped in free flight as they approached spider webs spun by *N. clavipes* positioned at the opposite end of the chamber. The webs, mounted on plastic rings, were fixed at predetermined distances from background patterns characterized by different spatial frequencies. The distance from which flies avoided webs in the different conditions was used to indicate the effects of background pattern on web perception.

The results of these experiments showed that both the type of background (ANOVA for categorical data; $P<0.01$; $n = 200$ flight paths for all of the laboratory analyses reported here) and the dis-

Table 4.3A Insect interception indices for experiments done with webs spun by adult *N. clavipes* and controls*

Background pattern: distance from web	Web	Control (no web)
1 mm: 2.5 cm	0.87	0.12
1 mm: 13 cm	0.84	0.15
1 mm: 2.5, 13 cm	0.22	0.08
3 mm: 2.5 cm	0.33	0.01
3 mm: 13 cm	0.14	0.03
3 mm: 2.5, 13 cm	0.29	0.07
5 mm: 2.5 cm	0.75	0.22
5 mm: 13 cm	0.34	0.13
5 mm: 2.5, 13 cm	0.1	0.06

*This ratio is defined to be: number of interceptions/manoeuvres through web plus web avoidances within 0.5 cm.

Table 4.3B Linear contrasts within web group for interceptions/avoidance

Stripe: distance versus stripe: distance	*F* value*	Significance
1 mm: 2.5 cm versus 1 mm: 13 cm	0.14	NS
1 mm: 2.5 cm versus 3 mm: 2.5 cm	26.9	P<0.005
1 mm: 13 cm versus 3 mm: 13 cm	46.5	P<0.005
1 mm 2.5, 13 cm versus rest	42.1	P<0.005
1 mm, 3 mm, 5 mm: 2.5 cm versus rest	39.6	P<0.005
1 mm, 3 mm, 5 mm: 13 cm versus rest	0.03	NS

F-value calculated using 0.026, error term of ANOVA for experimental groups only.

tance of the background from the web ($P<0.004$) affected the fly's avoidance behavior. As has already been mentioned earlier, more insects were intercepted at webs when backgrounds were close to the plane of the web than at a larger distance from it. In addition, flies were intercepted more frequently by webs backed by small-grain patterns than backed by large-grain patterns (table 4.3A and B). Furthermore, the analyses showed several very strong interaction effects indicating that the flies responded to the combined image of web and background. For example, there was no significant difference in the ability of flies to avoid webs placed at 2.5 cm or 13 cm from a background characterized by a high spatial frequency (1 mm black and white stripes, or a period less than or equal to fly body length). However, as the spatial frequency of the backgrounds was decreased (period of 3–5 mm black and white stripes, or three to four times fly body length), the webs apparently became more visible and were more successfully avoided by the flies. Hence, web site selection by spiders takes on a new meaning when considered together with the visual capabilities of insects. It is not only the mechanical properties of the web supports or type of vegetation, but also the frequency of the background pattern that might affect a spider's choice of a web site and thus the visibility of the web. In addition, the distance of the web from the background also plays a role. Interestingly, the finding that more insects were intercepted when the web was at a near distance (2.5 cm) from its background than when it was farther away (see above) was independent of the spatial frequency of the background.

Visible and invisible webs might have evolved in parallel

It may be that spiders have evolved strategies to vary web patterns in nonobvious ways that affect visual detection by insects. In some cases, it is more appropriate to consider spider webs as a type of visual display that actually attracts insects to them. In other cases, the web visibility may hinder prey capture if the insect recognizes the web as being dangerous. Although it is only when an insect flies into the web area that there is any chance of capture, insects that are close enough might see the web and avoid it. Therefore, with respect to insect-web encounter, some kind of a compromise must have taken place between the evolution of visible and invisible orb webs.

In the light of the diversity of orb designs and silk reflectance patterns, and the diversity of insects (large and small) that they intercept, the evolution of web patterns must have gone in several different directions. If the grain of an insect's ommatidia and its contrast sensitivity correlate with the species-specific patterns of orb webs, then the insect's limited visual resolution may be the operative reason why insects keep flying into webs after so many millions of years of evolution. On the other hand, when the insect's various visual capacities are coupled with its flight capabilities, it is difficult to understand how spiders catch any prey at all. In other words, the effect of silk reflectance and its consequences on the evolution of web design is likely to have been affected by the ability of insect prey to detect webs and avoid them.

Even though insects are found in habitats in which ambient light intensity varies over several log units, any individual species can survive in only a limited range of light intensities. For any given light condition, spiders need only to produce web patterns that are just slightly finer than the patterns a subset of insects can resolve from a particular distance. High reflectance resulting from the sparkling, bright droplets on the web spiral may attract insects from a larger distance. The link between the properties of insect spatial vision and the light conditions results in an "evolutionary space" within which aerial web patterns might have evolved. The insect color vision (see chapter 5) might have further expanded or limited the pathways through which spiders and their foraging strategies have evolved.

5

Insect Color Vision Is a Potential Selective Factor on the Evolution of Silk Chromatic Properties and Web Design

with M. Lehrer

Spiders produce multiple types of silk that vary in structural organization and reflectance properties, and therefore in their color. The limited amount of data available to date show that the physical properties of fibrous proteins produced by ancestral species differ from those produced by derived species. These variations might be correlated with the ambient light in which the spider forages (Craig 1997; but see Bond and Opell 1998 for results using a different analytic approach). The evolutionary success of the derived Orbiculareae could, at least in part, be related to shifts in silk reflectance. Insects, most of which are spiders' potential prey, are color-seeing animals. If it can be shown that shifts in silk color are correlated with spiders' foraging strategies and that they are tuned to insect color vision, then one could speculate that the evolutionary diversification of the reflectance properties of spider webs may be an adaptive response to the insects' perception of the spectral properties of the web's silk.

Most insects have UV-, blue-, and green-sensitive photoreceptors

Color vision evolved in insects and primates independently as a means to enhance the animal's ability to detect objects in its environment. The first condition that must be met for color vision to be possible is the existence of at least two spectral types of photoreceptor in the retina. The spectral properties of a photoreceptor are based on the visual pigment contained in it. Visual pigments have two components: a light-sensitive chromophore that is located in the retina and a protein (opsin) that forms a pocket in the cell membrane. Interactions between the transmembrane proteins and specific amino acid side groups of the chromophore determine the spectral sensitivity of the pigment. With only two types of chromophore, and a number of different opsins (Briscoe and Chittka 2001), several different visual pigments can be constructed. In some insects, a total of six different spectral classes of photoreceptor have been found; the common housefly, *Musca domestica*, has four (figure 5.1A) (Gogala 1967; Hardie 1986).

Most insects possess a trichromatic color-vision system that, based on phylogenetic and molecular analyses, dates back to the Devonian (Briscoe and Chittka 2001). In the most common cases, the three spectral types of photoreceptor are maximally sensitive to wavelengths in the ultraviolet (UV; 300–400 nm), the blue (400–500 nm), and the green portion of the spectrum (500–600 nm) (Peitsch et al. 1992). In the honeybee specifically, the maxima of these three receptor responses lie at 344 nm, 436 nm, and 556 nm, respectively (figure 5.1B) (Autrum and von Zwehl 1964; Menzel and Blakers 1976). There are a few deviations from this type of color vision. For example, crickets possess blue receptors only in the dorsal rim eye region, whereas in the remaining eye regions they have UV and green receptors (Labhart et al. 1984). Some butterflies and dragonflies have additional visual pigments sensitive to long-wavelength light (red, 600–700 nm) (Gleadall et al. 1989; Seki and Vogt 1998) (figure 5.2).

The response of a photoreceptor to monochromatic light (light of one particular wavelength) or to heterochromatic light (a combination of two or more wavelengths) depends on the fraction of the light that is actually absorbed by the visual pigment contained in that photoreceptor. The larger the amount of light absorbed, the higher is the light intensity monitored in the receptor involved, and the stronger is its response to the stimulus. The color system cannot discriminate between the color of a monochromatic light and the same color mediated by a mixture of several wavelengths. From light of a given wavelength or of a given combination of wavelengths, the different types of photoreceptor absorb different proportions of light (see figure 5.1). Thus, in a trichromatic color vision system, every light stimulus gives rise to three responses that are in a particular and unique ratio relative to one another. The color of the light stimulus is encoded in this ratio. It can be calculated from the sum of the three responses I_i (see also chapter 6), namely

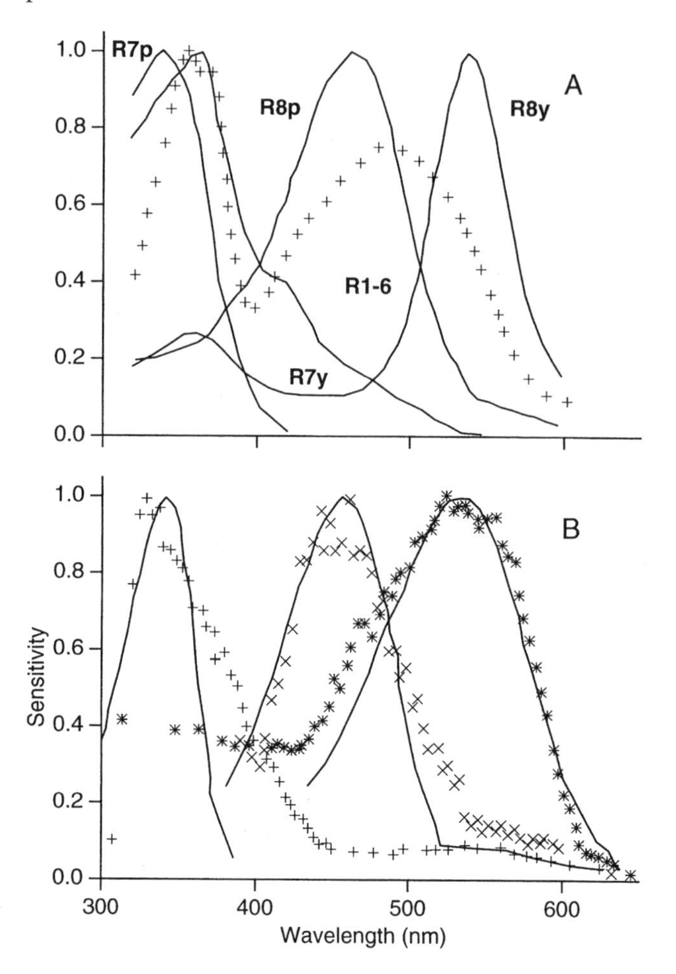

Figure 5.1 Spectral sensitivity functions of photoreceptor cells in the insect retina. The curves show the photoreceptor responses (ordinate) to monochromatic light of varying wavelengths (abscissa). Three spectral types of receptor, with response maxima in the ultraviolet (UV), blue, and green portion of the spectrum, are found in most insect species. (A) Spectral receptors of the housefly. (B) Spectral receptors of the honeybee (solid lines) and the stingless bee (x, +, *). (Redrawn from (A) Hardie 1986; (B) Peitsch et al. 1992; Menzel and Blaker 1976.)

$$I_i = I_{UV} + I_{bl} + I_{gr} \tag{1}$$

where I_{UV}, I_{bl} and I_{gr} denote the responses of the UV, blue, and green receptors, respectively. The sum I_i represents the total intensity of the stimulus under consideration, but tells nothing about its color. The three responses R_i evoked in the three receptor types (UV, bl and gr) are calculated as

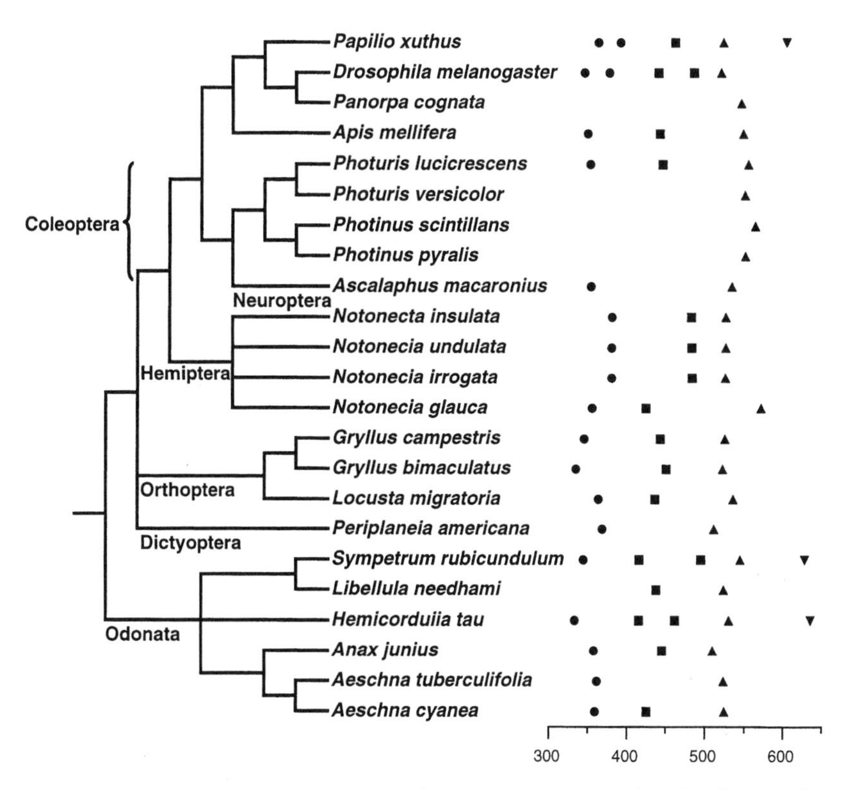

Figure 5.2 Spectral sensitivity of insect visual pigments superimposed on the diagram of their phylogeny. For each receptor type in each species, the wavelength of light that evokes the maximum response is shown. Circles: UV receptors; squares: blue receptors; upward triangles: green receptors; downward triangles: red receptors. (Redrawn from Briscoe and Chittka 2001.)

$$R_{UV} = I_{UV}/I_i \tag{2}$$

$$R_{bl} = I_{bl}/I_i$$

$$R_{gr} = I_{gr}/I_i$$

From these three values, the ratio R_{UV}:R_{bl}:R_{gr} can be calculated. Because this ratio unequivocally describes the color under consideration, it can be used to express every possible color as a point in a three-dimensional diagram, for example in the Maxwell chromaticity triangle (figure 5.3). Maxwell's triangle represents the so-called color space of the trichromatic animal under consideration, in our example, the honeybee. The three sides of the triangle are proportional to the three coordinates of the color space, each representing one of the three responses. All achromatic lights (white, black, and all shades of gray) share the white locus in the center of

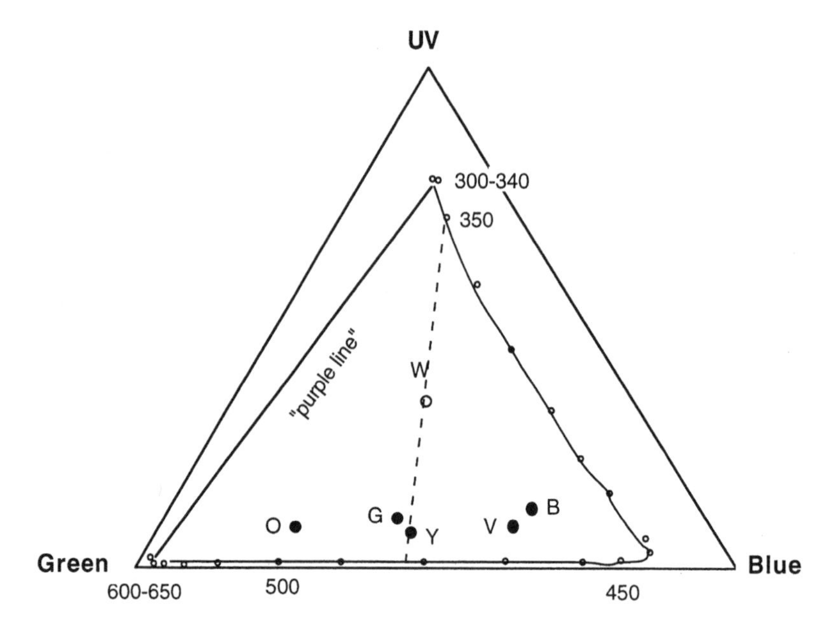

Figure 5.3 The honeybee's color space, represented in the so-called Maxwell chromaticity diagram. The three sides of the triangle provide the three coordinates of the relative responses evoked by each color stimulus in each of the three spectral types of photoreceptor. The ratio of the three responses defines the locus of every given color in the color space. The amount of color contrast between two colors is proportional to the distance between their loci. The greater this distance, the more do the two colors differ in the insect's eyes. The solid line connects the loci of monochromatic lights. (After Menzel et al. 1986.) *W* is the locus of white light. Monochromatic colors connected by a line passing through the white point (dashed line) are complementary. Loci of several heterochromatic colors are also shown. (After Lehrer 1999. Blue (B), violet (V), yellow (Y), orange (O), and green (G).)

the triangle (*W* in figure 5.3). They are defined as light that evokes equal responses in all three spectral types of receptor.

In some cases, a particular visual task is mediated by only one spectral type of receptor. For example, the perception of polarized light is mediated exclusively by the UV-sensitive receptor in both bees (von Helvesen and Edrich 1974) and ants (Wehner and Strasser 1985; Labhart 1986), and exclusively by the blue receptors in crickets (Labhart et al. 1984). In bees, the optomotor response is mediated exclusively by the green-sensitive receptor (Kaiser and Liske 1974), and so is the movement avoidance response (Srinivasan and Lehrer 1984), scanning behavior (Lehrer et al. 1985), and edge detection (Lehrer and Srinivasan 1993) (see chapter 4). In all these cases, the performance is color blind, because color vision requires the participation of more than one spectral type of receptor. Flies, which are color seeing in bright daylight (Menne and Spatz 1977; Fukushi 1989), are color-blind at night, because only one of their spectral receptor types (the green-sensitive receptors R1–6 in figure 5.1A) is active in dim light (Kirschfeld and Franceschini 1968). Many flying insects, including most Hymenoptera, are not active in dim light. In daylight, all insects are color seeing.

Detection of colored objects is based on their contrast against the background

Regardless of whether or not an object is colored, it can only be detected if it produces sufficient contrast against its background. This contrast may be either achromatic (that is, concerning only differences in intensity between object and background; see chapter 4) or chromatic (involving differences in color). In either case, higher contrast results in better detection. Thus, in the task of seeing objects in the environment, color vision is neither necessary nor sufficient. Its advantage over achromatic vision is that there are very many pairs of colors that produce contrast against each other, whereas intensity contrast varies along a one-dimensional scale.

Unlike the case of intensity contrast [see Eq. (1) in chapter 4], there is no equation that would allow calculation of the amount of color contrast produced by two adjacent color stimuli. Instead, the contrast between two colors is expressed by the distance of their loci in the insect's color space (see figure 5.3). Based on the principle underlying the construction of the color space, two colors that differ strongly in the ratio of the three responses come to lie farther apart than colors that differ less. Therefore, the greater the distance between two loci, the greater the color contrast that the two colors produce against each other.

Apart from the Maxwell diagram, several further models have been proposed for describing the insect color space using different methods for calculating color loci (Backhaus and Menzel 1987; Chittka 1992; Vorobiew and Brandt 1997). In principle, however, all methods allow for the same conclusions. For example, in the eyes of the honeybee, yellow differs from blue much more than from green, and blue differs from violet much less than from yellow (see figure 5.3), regardless of the model used for the calculation (see also Lehrer 1999). Thus, the color contrast that a yellow web produces against green foliage will be much lower than the contrast it would produce against blue sky. The highest color contrast is found between pairs of complementary monochromatic lights (see figure 5.3). In nature, however, light is never monochromatic; it is always a mixture of many wavelengths.

Lehrer and Bischof (1995) demonstrated that color contrast can replace intensity contrast in an object detection task. Using black discs of various sizes placed against a white background producing an intensity contrast of 83%, but no color contrast, the bees detected the disc even when it subtended no more than 3° at the eye. With a gray disc on a white background, producing intensity contrast of 39%, on the other hand, the disc was not detected unless it subtended more than 15° at the bee's eye. (For the calculation of intensity contrast, see chapter 4.) Thus, in the absence of color contrast, intensity contrast is the more effective the larger it is. More importantly, however, a blue disc placed in front of a yellow background was detected even when it subtended a visual angle as small as 5° at the eye, although the color combination used contained no intensity contrast. The same result was obtained using a violet disc that produced both color contrast and intensity contrast against the yellow background (Lehrer and Bischof 1995). Thus, intensity contrast is not necessary when color contrast is present.

These results show that the color vision system can accomplish even spatial tasks. Color contrast can replace brightness contrast not only in the object detection task, but also in tasks involving shape discrimination (Menzel and Lieke 1983; Lehrer et al. 1985; Lehrer 1997a; Lehrer 1997b; Lehrer 1999).

The perceptions of chromatic contrast and achromatic contrast are independent processes

Ever since the classical experiments conducted by Karl von Frisch (1915), it has been shown many times that stimulus intensity is irrelevant in color discrimination tasks (see for example Menzel and Backhaus 1989). Thus, the perception of intensity contrast and that of color contrast are two independent processes. In different behavioral contexts, insects may use either chromatic or achromatic contrast. In bees, for example, all motion-dependent responses investigated so far are governed by the green-sensitive receptor and are therefore based on achromatic contrast (for a review, see Lehrer 1997a,b). In the context of foraging, on the other hand, it is only color that counts (von Frisch 1915; Menzel and Backhaus 1989). And, finally, in object resolution tasks, it seems that the use of either chromatic or achromatic contrast depends on the distance of the object from the eye. In extending the experiments by Lehrer and Bischof (1995), Giurfa et al. (1996b) showed that small objects (subtending about 5° at the eye) are detected on the basis of achromatic contrast, whereas objects that subtend larger angles are recognized by their color.

Applying these findings to the question of web detection by insects, a yellow web is expected to be detected against the blue sky, regardless of brightness. The detection of a yellow web against green vegetation, on the other hand, will hardly occur on the basis of color contrast. Instead, it will depend on the amount of achromatic contrast between the two. The dark green color of old vegetation might render a bright yellow web visible, whereas the light green of young leaves that strongly reflect in the yellow portion of the spectrum will rather make the yellow web disappear. However, it may be that, analogous to the use of color in the French painting school of pointillism, a yellow web suspended against dark green vegetation might make the vegetation behind it appear brighter, thus being suggestive of young green leaves that might attract some herbivorous insects (see chapter 6).

The question indeed is whether the color of webs has, in the course of evolution, been affected by the color vision system of insects. To address this question, we must look at the spectral composition of the various spider silks.

Webs of ancestral and derived spiders differ in their spectral reflectance

The color of silk, particularly that of ancestral spiders, is not based on the presence of a pigment, but rather on the structural organization of the protein involved, for example of the configuration of the protein (see chapter 3). Therefore, any evolutionary process that might have caused a shift in silk color must be correlated with new protein structures (Craig et al. 1994). As far as it is known, fibrous proteins and silks spun by ancestral mygalomorph spiders are similar in brightness and have important common spectral properties, although various species differ in the type of web they spin.

Spectral reflectance of silk spun by ancestral spiders is characterized by a pronounced peak in the UV region

The phylogenetically primitive spiders of the family Mygalomorphae produce silks that are similar with respect to the peak in UV (at 350 nm), but differ from one another in the reflectance at wavelengths greater than 400 nm. For example, silks spun by *Aphonopelma* sp. (Theraphosidae) and *Ischnothele* sp. (Dipluridae), both of which reflect strongly in the UV, differ in the proportion of yellow and green from silks spun by *Hexurella* sp. (Mecicobothriidae) (figure 5.4A). The former two range from 60% to 80% reflectance between 350 nm and 400 nm, and show a smaller, broader secondary peak in the yellow-green region of the spectrum at 550–650 nm. Silks spun by *Hexurella* sp. are, again, characterized by a peak in the UV (at 370 nm), but reflectance drops to about 45% at all other wavelengths (see figure 5.4A). Webs of some species (for example *Diguetia* and *Sosippus*) reflect strongly at all wavelengths, including UV (see figure

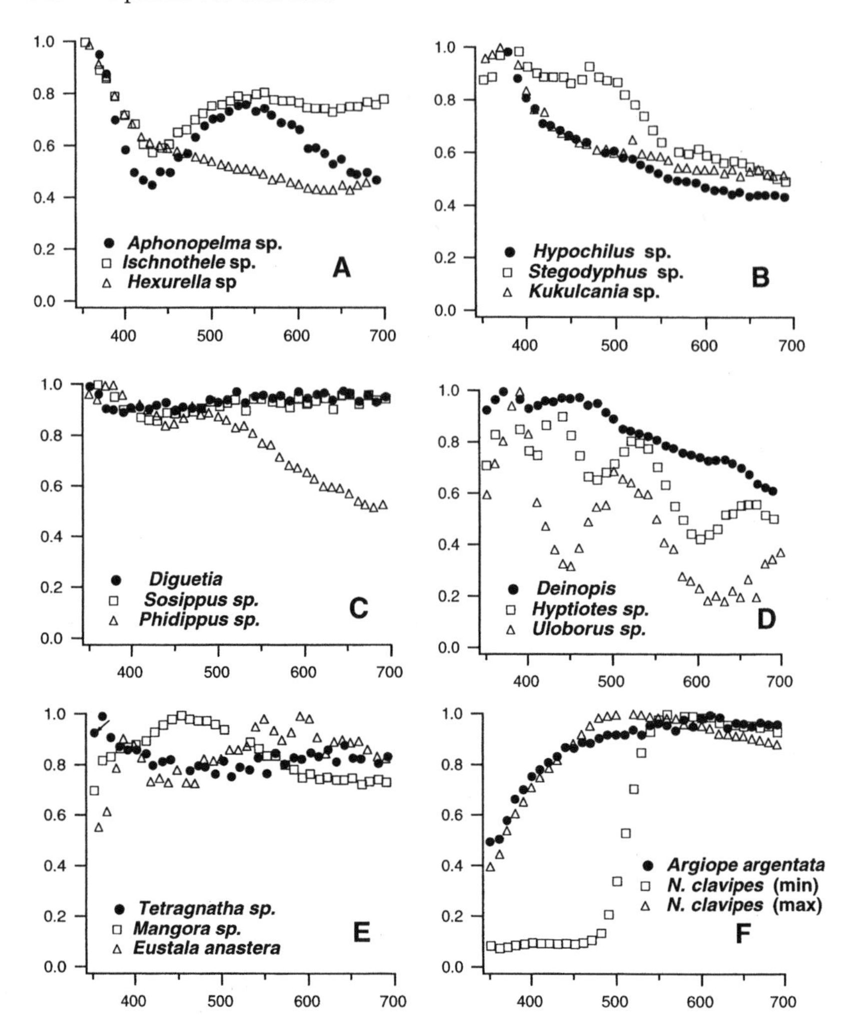

Figure 5.4 Spectral properties of silk spun by spiders in the Mygalomorphae (A) and the Araneomorphae (B–F) with cribellate (B, C) and ecribellate (D–F) spiders, primitive aerial web weavers (C), and derived aerial web weavers (E, F). (A) Silks spun by mygalomorph spiders *Aphonopelma* sp., *Ishnothele* sp., and *Hexurella* sp. (B) Silks spun by cribellate, araneomorph spiders, *Hypochilus* sp., *Stegodyphus*, sp., *Kukulcania* sp. (C) Silks spun by ecribellate, non-web-weaving spiders *Phidippus*, and sheet web weaving *Sosippus* and *Diguetia*. (D) Silks spun by primitive, cribellate aerial orb web weavers, *Deinopois*, *Hyptiotes*, *Uloborus*. (E) Silks spun by ecribellate, derived aerial-web weavers. (F) Silks spun by *Nephila clavipes* (maximum reflectance achieved with unpigmented silk), *Nephila clavipes* (minimum reflectance achieved with yellow, pigmented silk) and *Argiope argentata*. (Redrawn from Craig and Bernard 1990; Craig et al. 1994.)

5.4C), and must therefore appear white to the insect. Thus, all silks that have been measured so far and that are spun by mygalomorph spiders (see chapter 3) are characterized by a spectral peak in the UV region of the spectrum (figure 5.4A), whereas silks spun by some cribellate Araneomorphae are characterized by a spectral peak in the UV and others are characterized by a flat reflectance band that includes both the UV and blue (table 5.1).

The silks spun by *Hypochilus* (Hypochilidae), a member of the most primitive family among the araneomorph spiders (Platnick 1977), and its related taxa, *Kukulcania* (Filistatidae) are again characterized by a spectral peak in the UV region (figure 5.4B). Reflectance drops to 40–60% at wavelengths between 550 nm and 700 nm. The spectral properties of all silks spun by the Uloboridae measured to date have a well-defined peak in the UV (350–400 nm) which lies at least 20–30% above the reflectance at other wavelengths and, depending on the species, one or two secondary peaks at longer wavelengths (figure 5.4D). The spectral reflectance of the silk spun by *Deinopis* (Deinopidae) (figure 5.4D) is very similar to that of the nonaraneoid spider *Stegodyphus* (Lycosidae) (figure 5.4B). The peak in the UV region is accompanied by a shoulder in the blue region of the spectrum. Again all these spiders forage, in dim or nocturnal environments (see table 5.1). Whereas most of the silks spun by the nonorbicullate, araneomorph spiders reflect UV light (see above), two genera—*Diguetia* (Diguetidae; nocturnal, sheet web) and *Sosippus* (Lycosidae; nocturnal, sheet web)—produce spectrally flat silks (figure 5.4C). Both these species are nocturnal. *Phidippus otiosus* (Salticidae), on the other hand, forages in bright light and produces silks with high reflectance in the UV and blue regions, and lower reflectance in the green region of the spectrum (figure 5.4C). *Phidippus otiosus*, however, does not weave a prey capture web. Thus, web-weaving ancestral spiders have two features in common: all spin silk with a pronounced reflection in the UV portion of the spectrum, and all forage at night or in dim light conditions.

Whereas webs of the ancestral spiders are homogenous with respect to UV reflection (see above), they are rather diverse with respect to web construction (table 5.1). Within all of these genera, silks are used in several quite different ways. The phylogenetically primitive aerial web weavers are grouped into the superfamily Deinopoidea that includes twenty-three genera in two families. All of the spiders in the family Deinopidae (four genera, fifty-six species) weave the same type of web. The Uloboridae (19 genera, 242 species), on the other hand, weave a variety of webs, the most ancestral of which are orb-webs. Hypochilidae, the most primitive family within the araneomorph spiders, weave sheet webs—and so do spiders of the families Diguetidae and Lycosidae. Finally, some species

Table 5.1 Silk spectral properties, light habitat, and foraging modes (Craig 1994; Bond and Opell 1998), listed for several families, from ancestral Mygalomorphae to the derived Araeomorphae. Data on web type and ambient light are from Shear (1986), or based on observations made during silk collection on several occasions.

Status	Family/superfamily		Light habitat	Spectra	Web type
Mygalomorphae					
	Theraphosidae	*Aphonopelma* sp.	Nocturnal	UV peak	None
	Dipluridae	*Ishnothele* sp.	Nocturnal	UV peak	None
	Mecicobothridae	*Hexurella* sp.	Nocturnal	UV peak	Funnel and Sheet
Ancestral araneomorphae					
Paleocribellatae	Hypochilidae	*Hypochilus s* sp.	Dim	UV peak	Vertical sheet
Haplogynae	Filistatidae	*Kukulcania*	Nocturnal	UV broad	Sheet
Entelegynae (Derived araneomorphae)					
	Eresidae	*Stegodyphus* sp.	Nocturnal	High reflectance at short wavelengths (<550 nm)	Funnel and Sh
	Lycosidae	*Sosippus* sp.	Nocturnal	Flat	Sheet
	Diguetidae	*Diguetia* sp.	Nocturnal	Flat	Sheet
	Salticidae	*Phidippus otiosus*	Bright	High reflectance at short wavelengths (<550 nm)	None
	Deinopoidea Deinopidae	*Deinopis* sp.	Nocturnal	UV broad	Modified orb
	Uloboridae	*Miagrammopes animotus*	Nocturnal	Peaks in UV, blue, green, orange	Space
	Uloboridae	*Uloborus glomosus*	Dim	Peaks in UV, blue green	Reduced orb
	Uloboridae	*Hyptiotes cavatus*	Dim	Peaks in UV, violet, green, red	Reduced orb

Uloboridae	*Philoponella tingens*	Dim	Peaks in UV, violet, blue, orange	Orb
Araneoidea Theridiidae	*Latrodectus mactans*	Dim	Flat	Space
Theridiosomatidae	*Epilineutes globosus*	Dim	Low reflectance at short wavelengths (<390 nm)	Orb
Tetragnathidae	*Leucauge* sp.	Bright, dim	Flat	Orb
Tetragnathidae	*Tetragnatha* sp.	Bright, dim	Flat	Orb
Tetragnathidae	*Nephila clavipes*	Bright, dim	Low reflectance at short wavelengths (<390 nm)	Orb
Araneidae	*Eustala* sp.	Nocturnal	Flat	Orb
Araneidae	*Eustala anastera*	Nocturnal	Flat	Orb
Araneidae	*Neoscona domiciliorum*	Nocturnal	Flat	Orb
Araneidae	*Mangora pia*	Dim	Flat	Orb
Araneidae	*Argiope argentata*	Bright	Low at reflectance short wavelengths (<350 nm)	Orb
Araneidae	*Micrathena schreibersi*	Bright	Low reflectance at short	Orb

such as *Phidippus otiosus* do not weave webs at all (see table 5.1). Thus, although the webs of ancestral spiders are similar with respect to their content in the UV reflectance, they are diverse with respect to web type (if any) and construction.

The spectral reflectance of silk produced by derived spiders is characterized by the absence of a peak in the UV region

The ecribellate spiders, those that do not weave webs (figure 5.4C), and those that do (figure 5.4E and F), produce silks that are spectrally flat or are characterized by reduced reflectance in the UV region. These spiders forage both at night and in a variety of diurnal conditions.

All of the Araneoidea, the derived orb-weavers, are ecribellate spiders. Their silks differ from those of the ancestral, cribellate, orb weavers, the Deinopoidea, in both spectral properties and the mechanism by which their webs retain prey. None of the araneoid silks is characterized by a spectral peak in the UV, and none of them retains prey via brushed fibrils (see chapter 3). Some of the derived web weavers, such as *Eustala anastera*, *Eustala* sp., *Mangora pia*, *Leucauge* sp., *Tetragnatha* sp., *Neoscona domiciliorum*, and *Latrodectus mactans*, produce silks with a flat reflection curve that therefore appear white. All these spiders forage at night or in dim light (table 5.1). Alternatively, the diurnal araneoids that spin silks characterized by reduced reflectance in the UV region forage in either bright or dim light. These silks include those spun by the diurnal spiders *Nephila clavipes*, which forages in enclosed forest understory as well as light gaps and forest edge, by *Argiope argentata*, which forages in bright light (figure 5.4F), and by *Epilineutes globosus*, which forages in dim light (table 5.1). The silks spun by all of these spiders reflect only little in the UV (up to 370 nm) and are therefore all likely to appear blue-green to the insect eye. Of the silks measured to date, all those that lack a peak in the UV region of the spectrum are spun by spiders that forage in open or bright light habitats (figure 5.5). None of the derived orb weavers spin silks that reflect UV. Therefore, any light that is reflected by their silks is expected to produce contrast against every background that reflects UV.

In summary, the derived spiders differ from ancestral spiders in several respects. They are diverse in their ways of life, some being nocturnal, others diurnal, whereas ancestral spiders are all nocturnal. The manner in which a nocturnal or a diurnal way of life is affected by the light conditions under which the spiders forage is obvious. Further, none of the araneoid spiders produces silks with a pronounced UV reflectance peak that characterizes the ancestral Orbiculareae, the Deinopoidea.

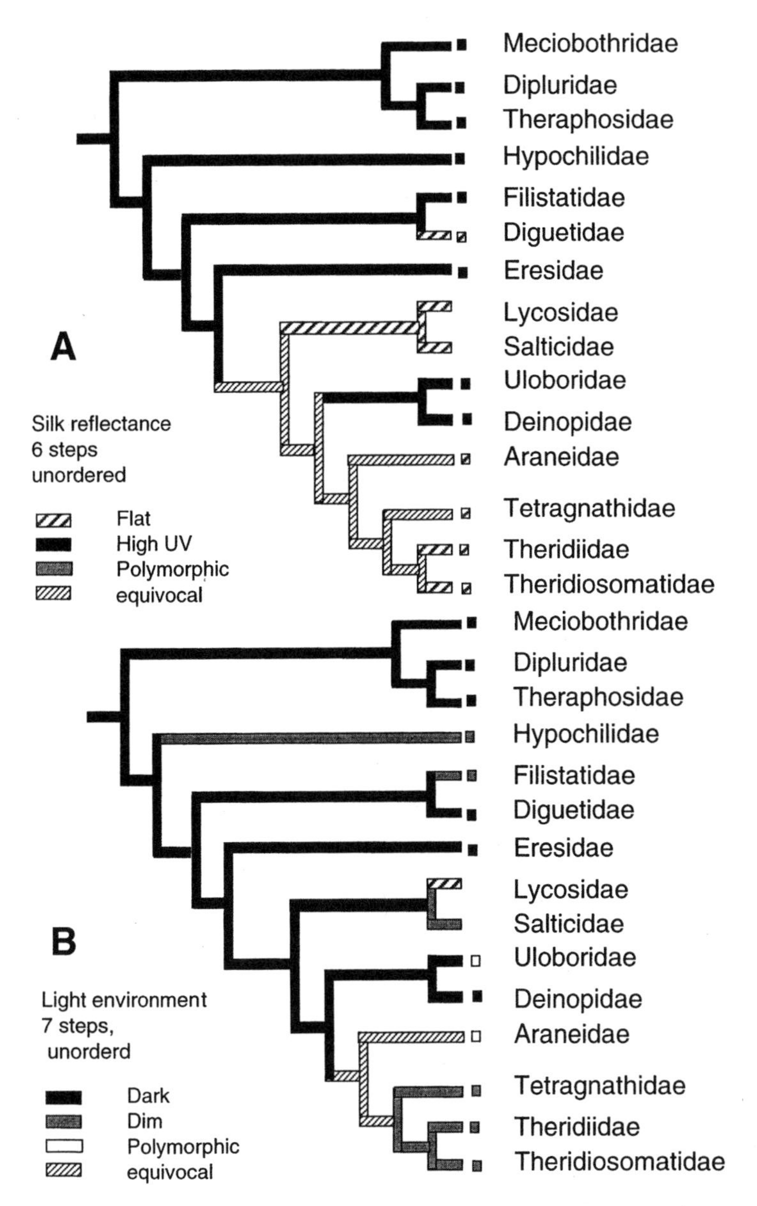

Figure 5.5 Spectral properties of silks and spider foraging environments mapped onto a cladogram of the Araneae. (A) High UV reflectance is ancestral for the Araneomorphae and for the Orbiculareae specifically, spiders in the superfamily Deinopoidea (Deinopidae and Uloboridae). Low reflection in the UV, or a flat reflectance curve are unequivocally characteristic of the dragline and capture silks spun by the Araneoidea. (B) Ancestral orb spiders that spin UV reflecting silks forage in dark or dim light. The derived orb weavers, the Araneoidea, forage in a greater diversity of light conditions and produce silks whose spectral properties are more diverse than those of their closest ancestors, the Deinopoidea. (Redrawn from Bond and Opell 1998. Composite tree; support values for the nodes not available; J. Coddington, J. Bond pers. commun.)

Do the differences in UV-reflection of silks result from selective effects of insect color vision?

All insects, regardless of other details of their visual systems, detect UV light. Because the only natural source of light is skylight (sunlight), light that contains UV might act as a signal. For example, insects searching for open space are attracted to UV light, whereas those attempting to find enclosed or protected areas may turn away from it. However, a given surface can only reflect in the spectral region of the light that is incident on it. Because there is no skylight illumination and therefore little if any incident UV light at night, the UV peak of the reflectance cannot be of any advantage to nocturnal spiders. It must simply be the consequence of protein structure that has not been selected against in the course of evolution. As has been emphasized in an excellent recent review (Briscoe and Chittka 2001), one cannot take every observed feature to be a consequence of selection, or based on adaptation to the environmental conditions. Some features might have remained conserved not because they are particularly advantageous in a given habitat, but rather because they are not particularly disadvantageous to the animals carrying them.

If there is any effect at all of insect color vision on the evolution of silk color, then it can only be expected to have acted in the diurnal spider species, whose prey are, naturally, diurnal insects. The pronounced reflection in UV has disappeared from the silks spun by derived orb weavers. Depending on the contrast that the webs produce against their background, the webs will either be invisible (thus deceiving the insects), or will be visible (with a potential of being attractive to insects). For example, a yellowish web will be invisible against green foliage, but might be conspicuous against the blue sky. A bluish web, on the other hand, will be invisible against the sky, but visible against foliage. Finally, in cases in which the web retains a sticky spiral with shiny, sparkling viscid droplets, the web is visible and is likely to attract insects (see chapter 4). However, it is only the insect's behavior that can show whether a web is visible or invisible, attractive, or repelling.

In the next chapter we will therefore look at insects' behavioral responses to several parameters of spider webs by presenting them with different silks and web patterns in natural, semi-natural, and controlled laboratory settings.

6

Insect Learning Capacity Is a Potential Selective Factor in the Evolution of Silk Color and the Decorative Silk Patterns Spun by Spiders

The two previous chapters have discussed the spatial and chromatic visual capacities of insects, to outline how—or if at all—these capacities might have acted as a selective force on the evolution of silks. The previous findings suggest two pathways through which insects may have affected silk evolution. One pathway results in the evolution of silks that are either cryptic against their background or invisible to insects; the second pathway results in silks that display some spatial or chromatic parameters that attract insects to the web.

At first sight, web attractivity appears to be advantageous to the spider by increasing the probability of prey capture. However, it might have the opposite effect if the prey are armed with a learning capacity equal to that demonstrated for several insect species in behavioral experiments. An insect that has been attracted by a web and intercepted will be badly punished. If the insect succeeds to struggle free, then the memory of the annoying experience might cause it to avoid webs in the future. This avoidance behavior would select against the evolution of visible webs and favor the evolution of cryptic webs.

In the present chapter, we will review several behavioral studies designed to examine how insects respond to natural webs under a variety of experimentally manipulated conditions. First, we will determine the particular conditions in which webs are attractive to insects. Mainly, however, we will examine the insects' capability to develop an avoidance response toward spider webs based on previous experience at a web.

Insect responses to visual cues are either innate or learned

Most experiments on color and shape learning in insects have been conducted on honeybees, using a training method in which the insect is rewarded with sucrose solution upon choosing the stimulus to be learned. The bees' performance in associating food with a visual stimulus is excellent. Many colors render more than 70% correct choices after only one learning trial, and all colors are chosen correctly by more than 80% of the bees after only three rewarded visits (Menzel 1967; Menzel et al. 1974). Shapes require longer training periods (ten to thirty rewarded visits, depending on the shape), but are nevertheless learned very well (often to more than 80% correct choices; Wehner 1981; Lehrer et al. 1985). Associative learning through reward has been demonstrated in several other groups of insects, such as butterflies (Lewis and Lipani 1990; Weiss 1995; Kelber and Pfaff 1999), locusts (Raubenheimer and Tucker 1997), moths (Kelber 1996; Kelber 1997), and even flies (Nelson 1971; Fukushi 1976; Fukushi 1989).

In many cases, a preference for a particular feature of a stimulus need not be learned. It is genetically fixed and therefore spontaneous. Such innate preferences are likely to be based on the features in the animal's natural habitat that have yielded rewards over evolutionary time. For example, without any training procedure, bees prefer colorful, small, and high-contrast targets, targets of high spatial frequency (see chapter 4) as well as radiating shapes (Lehrer et al. 1995; Lehrer 1997b) and symmetrical patterns (Lehrer et al. 1995; Giurfa et al. 1995). All of these features are typical of natural flowers. Inexperienced foragers express a spontaneous preference for particular colors (Giurfa et al. 1996b); however that is easily erased by later experience with other colors (Menzel 1967; Giurfa et al. 1995).

Flies, in contrast to bees, search for food randomly and therefore they need not remember the features of a once-found food source. Associative learning in flies (see references above) is slow and the training difficult. However, flies learn rapidly via aversive training methods, that is, by punishing the insect for making a wrong choice. Aversive learning of colors was achieved by Menne and Spatz (1977) by applying a shaking procedure to punish the flies, and learning of shapes by punishing the flies with heat (Ernst and Heisenberg 1999). In the bee, color and shape training experiments using aversive methods have never been reported. It is conceivable that bees fail to associate spectral and spatial cues with an aversive experience because these cues are being innately associated with a food reward. Previous training experiments on bees have used electroshocks, conducted in the context of responses to the earth's magnetic field (Walker and Bitterman 1989) and air movement (Towne and Kirchner 1989); two cues that, in nature, never predict a reward.

Avoidance responses also occur spontaneously. One only needs to think of the escape responses of insects observed in the field. If a particular visual cue always predicts a dangerous experience, the animal is very likely to evolve a genetically fixed avoidance response towards that stimulus. Previous experiments on avoidance learning have been carried out in the laboratory using punishment procedures that do not occur in nature and that are not related to any particular environmental conditions. In the experiments described below, both the behavioral context (foraging) and the punishment (interception) are natural and realistic.

Some silks possess particular spectral or spatial features that vary with ambient light

In most cases, silk color is the result of the structural properties of the protein. However, there are some spiders that produce pigments that mask the structural color of the silk, thus effectively modifying its spectral properties. For example, *Nephila clavipes* adds yellow pigments to its silks and seems to vary silk color depending on the ambient light intensity in which the spider is found (figure 6.1). The pigments are made up of two quinones, xanthuric acid, and a fourth, undetermined compound (Holl and Henze 1988). The colors of silks, and hence of the webs, might affect the insects' response to them. If the web possesses a spectral reflection that is similar to that of the background, then it will produce no color contrast and might be difficult to see (see chapter 5). On the other hand, with a different reflection spectrum, or against a different background, the web might be visible and could act to either attract insects or to repel them.

Pigment synthesis is likely to cause the spider considerable metabolic costs and is therefore expected to have some useful function. Perhaps attesting to this is the fact that most other spiders do not actively vary the color of their silks. However, most spiders are not found in the wide range of habitats occupied by *Nephila*, nor do they spin the very large webs and web threads that are correlated with *Nephila*'s large size. *Nephila* sp. forage in tropical and subtropical sites, in the forest understory and in nonforest sites (Robinson 1973; Higgins and Buskirk 1992; Miyashita 1990; Miyashita 1992). It is possible that pigments enhance the foraging performance of *Nephila*, but that spiders have evolved to produce pigments only in sites where the web is expected to produce color contrast against the natural background.

N. clavipes varies web color in response to the total amount of light in its habitat, and to the spectral composition of the ambient light, suggesting that silk color may serve an adaptive function. To examine the correlation between web color and ambient light under

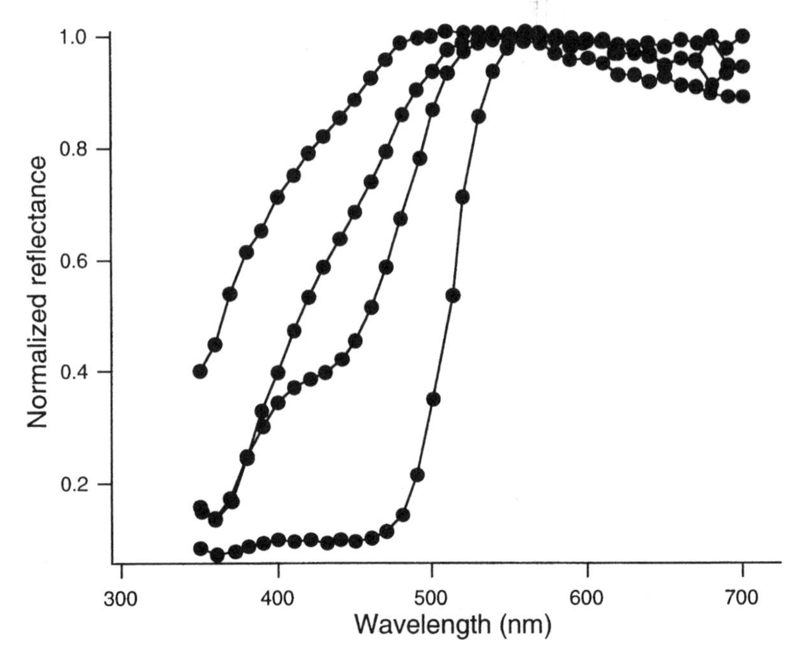

Figure 6.1 Spectral properties of viscid silks collected from webs spun by freely foraging *Nephila clavipes. Nephila clavipes* silks are characterized by a range of spectral properties. However, all of them appear colored (not white) to the insect eye, because they lack reflectance in the short-wavelength range of the spectrum. To the human eye, these silks possess various hues of yellow. (Redrawn From Craig et al. 1996.)

controlled conditions, *N. clavipes* were maintained in an experimental chamber under two different light intensities, one half as high as the other (figure 6.2). The spectral reflectance of the silks produced by the spiders was found not to differ between the two light conditions, but that the total amount of pigment the spiders produced depended on the ambient light intensity. In bright light, the spiders produced more pigment than in dim light (Craig et al. 1996; figure 6.3). In natural conditions, the intense golden color of the silk will result in high color contrast when the web is viewed against the blue sky. If the web is viewed against bright green, background vegetation, contrast is low. The particular features of the web site and angle at from which the web is viewed will determine web contrast and potential visibility to prey.

N. clavipes also seems to vary natural web colors according to the spectral composition of their ambient light environment. For example, the spectral spectrum (the distribution of light) of forest understory on Barro Colorado Island (BCI; field site of the Smithsonian Tropical Research Institute, STRI) is rich in wavelengths between 400 and 500 nm (blue) (Endler 1988) (figure 6.3A). Spiders maintained in an experimental chamber in light conditions similar to

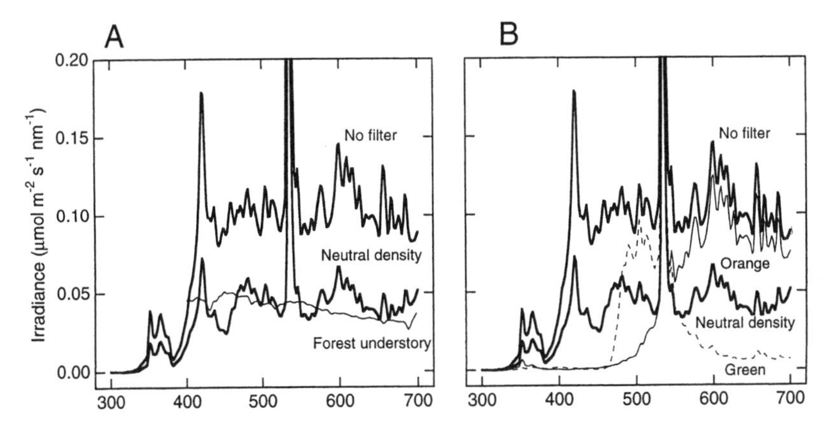

Figure 6.2 Spectral composition of light of two different intensities used in the experimental chamber, and transmission properties of the filters used. (A) Spiders were maintained in bright light or in light reduced by 50% using neutral filters. The reduced light in the chamber was similar in intensity and spectral content to light in the forest understory on Barro Colorado Island (data from Endler 1993). (B) Two additional filters were used to manipulate the spectral content of the light. An orange filter eliminated all wavelengths shorter than 500 nm, and transmitted almost all wavelengths greater than 540 nm; a green band-pass filter transmitted light between 480 and 550 nm but eliminated all other wavelengths. (From Craig et al. 1996.)

those in forest understory (480–550 nm) produced masking compounds that absorbed very little light at wavelengths longer than 400 nm. Thus, both in the forest understory and in dim laboratory conditions, spiders produce silks that would appear to be pale bluish-green, or pale yellow to the insect eye. In nonforest habitats, as well as in forest light gaps however, the ambient light is rich in wavelengths between 500 and 650 nm (yellow and orange). When the light in the chamber was adjusted to be similar to light in nonforest sites, the spiders produced yellow and yellow-orange pigments (figure 6.3B). Specifically, the silks reflected wavelengths longer than 500 nm, but absorbed wavelengths shorter than that. In general, the pigmented webs reflected less light than unpigmented webs, and were similar in color to silks collected from *N. clavipes* that foraged naturally in gaps or at the forest edge. During all of these laboratory experiments, food type and quantity were kept constant. Hence, the differences in pigment properties could not have been the result of diet. The data therefore show that the spiders were sensitive not only to light intensity but also to its spectral composition. Furthermore, the spiders responded to each of these two parameters independently from the other (Craig et al. 1996).

Adjusting the amount of pigment to the ambient light intensity might have the function of preventing the occurrence of contrast between the web and the background. In most cases, the background consists of natural vegetation, reflecting all wavelengths longer than

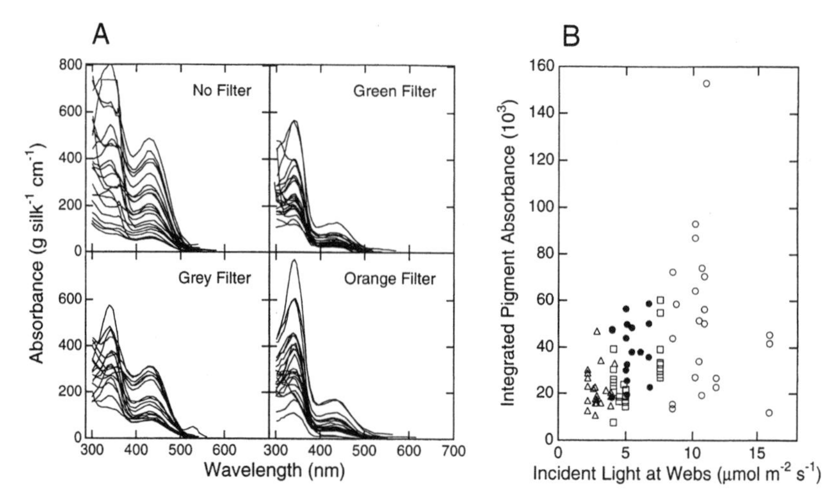

Figure 6.3 The absorption spectra of pigments. (A) Spiders maintained under light characterized by a flat spectral distribution but two different intensities (with and without a neutral filter) produced pigments that absorbed wavelengths shorter than 400 nm and reflected wavelengths longer than 500 nm. Spiders maintained in light from which some wavelengths were excluded (using an orange filter or a green filter) produced pigments that reflected most wavelengths longer than 400 nm but absorbed most wavelengths shorter than 400 nm. (B) Spiders maintained in light of a similar wavelength distributions but at either high (empty circles, no filter) or low intensity (filled circles, gray filter) produced pigments that differed in their spectral absorption properties proportionally to light intensity. Spiders maintained in light of different spectral distributions (gray filter: filled circles, orange filter: squares) but similar intensity produced pigments that absorbed the same total amount of light, but differed in their spectral absorption properties. Spiders maintained at the lowest light intensities (green filter, triangles) produced the smallest amount of light-absorbing pigments. (From Craig et al. 1996.)

about 500 nm. Pigmented silks possess a similar reflectance (see figure 6.3) and will therefore produce no color contrast against the vegetation. However, the silk might produce brightness contrast if it reflects less light than does the background. At high light intensities, more light is reflected from the vegetation. Because this light is mainly long-wavelength light, the web will appear darker than the background unless it, too, reflects more long-wavelength light. This is achieved by increasing the amount of yellow pigment.

In unpigmented webs, such a regulation is not possible. Unpigmented silks of ancestral spiders reflect ultraviolet (UV) light (see chapter 5; figure 5.4A–E). They will therefore produce color contrast as well as brightness contrast against vegetation. Insects might thus easily detect and avoid them. Unpigmented silks of more advanced spiders, on the other hand, reflect in the same spectral region as does the vegetation (see chapter 5; figure 5.4F). The webs of these spiders would therefore produce only little color contrast against the background vegetation. In bright daylight, they will probably produce no intensity contrast either, but they might appear brighter than the background in dim shadowy forest sites or in twilight, especially if the illumination angle is oblique to the web surface.

The finding that invisible webs, whether pigmented or not, are characteristic of mainly more advanced spiders supports the idea that cryptic webs might have evolved under the influence of insects' avoidance response.

Some silks and webs possess particular spectral or spatial features that might be attractive to insects

As discussed in chapter 5, achromatic contrast is more effective the higher it is, and chromatic contrast is effective even in the absence of intensity contrast (Lehrer and Bischof 1995). Silk colors will be visible to insects only in environments that are bright enough, and only if the web and its background differ sufficiently in either brightness or color. Of course, whether or not a site is sufficiently bright to be seen, depends on the visual system of the insect. For example, a forest-dwelling insect with a dark-adapted receptor system may be able to see silk colors at relatively low light intensities, whereas nonforest insects with a bright-adapted receptor system may not. An insect that makes use of both forest and nonforest habitats needs a visual system that functions under both light conditions. Flies, for example, possess a scotopic system consisting of one particular receptor type that is active in dim light but saturates in bright light, and a photopic system that only becomes active in bright light (see chapters 4 and 5).

We set out to tease apart the effects of color, pattern, and light by studying how stingless bees, *Trigona fulviventris* (Meliponidae), responded to three different types of web in four different forest sites. We used natural webs of *N. clavipes* and *A. argentata*. Webs of *N. clavipes* were either heavily pigmented (reflecting less than 10% of the light between 390 nm and 480 nm, but nearly 100% of wavelengths above 550 nm), or only slightly pigmented (see previous section).

A. argentata, a nonforest spider, produces unpigmented silks, the spectral properties of which are similar to those spun by shade-dwelling *N. clavipes* (figure 6.4). In fact, a comparison between silks of the two genera shows only a 5–10% difference in reflectance at wavelengths greater than 420 nm. Therefore, as in the case of the unpigmented webs spun by *Nephila*, the *A. argentata* webs would appear blue-green to the bee. They are, however, much less densely woven than the webs of *N. clavipes* (figure 6.6C). Therefore, even though *A. argentata* webs reflect the same wavelengths as some webs spun by *N. clavipes*, the total amount of light reflected from *A. argentata* webs is much smaller than that reflected from webs of *N. clavipes*. However, it is possible that the yellow pigments reduce the total amount of light

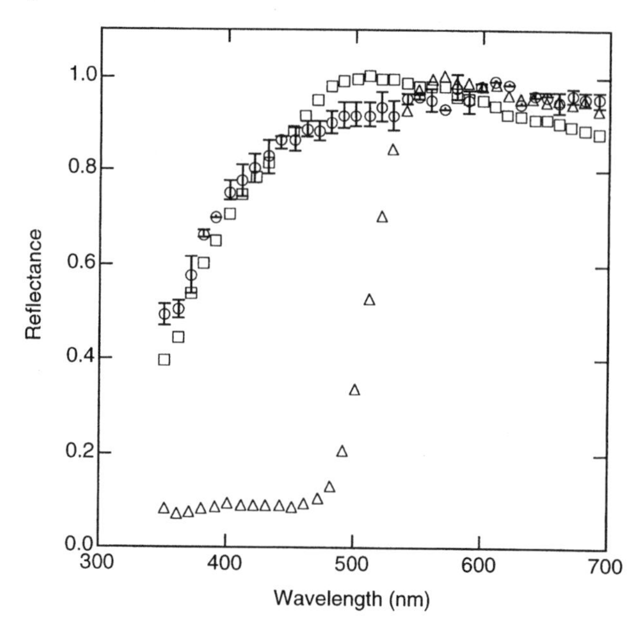

Figure 6.4 Contrasting the spectral properties of viscid silks and webs produced by *A. argentata* and *N. clavipes*. Spectral reflectance of silks produced by *N. clavipes* vary in yellow hue (maximum pigment, empty triangles; no pigment, empty squares) and hence affect web visibility against to its background. For *A. argentata* (empty circles), the average and SD of reflectance spectra of silks spun by four individuals are shown. They reveal only small variations in color. All samples were collected from webs spun by adult females, and all individuals of each species were approximately of the same size. (From Craig et al 1996.)

reflected from *Nephila* webs and thus counteract its higher total brightness that is due to the web's dense architecture (figure 6.6A,B). In this case, the webs of the two species may reflect comparable total amounts of light despite the differences in web density. The experiments were designed to test this possibility (Craig 1994a).

The stingless bee, *T. fulviventris*, used in these experiments builds its colony in the forest, but it forages in both forest and nonforest sites. We trained bees from four different nests to forage at four different artificial feeding sites, using the three types of webs described above. Thus, a total of twelve different experiments (three per hive) were conducted with bees from the four hives. Two of the experimental sites were in locations where the canopy above them was open, and two were located where the canopy was closed. A feeding dish containing a 0.25 M sucrose solution was placed at each site and an empty, black ring was placed vertically in front of it so that the bees had to fly through the ring to reach the food (figure 6.5). Two sides of the dish were blocked by vegetation; a third side was left open. During the training periods, although bees had access to the dish from two sides (the front side with the ring, and side 3), they usually flew on a bee-line through the open ring to reach the dish. During the test, the

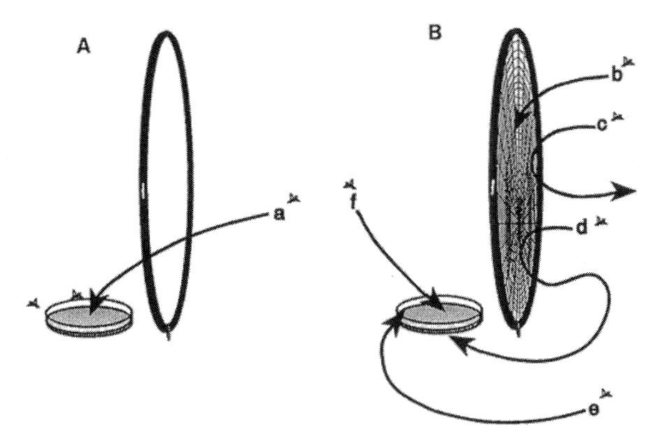

Figure 6.5 Observed responses of bees arriving at the experimental site. All bees were trained to fly through a training ring (a) to reach a dish containing sugar water. When confronted with a spider web, bees were either (b) intercepted, or (c) they retreated, approached the web (d) and flew around it, bypassed the web (e), or (f) approached the dish directly from the back. For each test, the proportion of intercepted bees was calculated from the total number of bees that arrived at the ring. (Reprinted with permission from *Behavioral Ecology and Sociobiology,* Volume 35 © 1994, Springer-Verlag. "Predator foraging behavior in response to perception and learning by its prey: interactions between orb-spinning spiders and stingless bees." Pp. 45–53. Figure 2A,B.)

empty ring was replaced by a ring bearing a web. No spiders were present so that captured bees always escaped. Upon arrival at the web, the bees were either intercepted, in which case they always struggled free, or approached the web, scanned its surface, and retreated, or reached the feeding dish by flying around the web or approaching the dish from the back (figure 6.5).

After two days of trials, the web was replaced with an empty ring to test whether the bees had been responding to the web, or just to the site at which the web was presented. The results of these tests (not illustrated) showed that the web, and not the site, was crucial to the bees' response. After two hours, most of the bees were flying through the ring. The next day their behavior was indistinguishable from that on the first day (control) of the experiments.

The statistical analysis of these data (table 6.1) revealed that the bees' response was strongest in the first 30 minutes of the experiment. Furthermore, they responded differentially to each of the three web types over the three-day period. The details of the web's site, such as background vegetation and ambient light intensity, affected the visibility of the web significantly. When bees encountered webs characterized by different fiber densities but spun from similar silks, it was found that the bees avoided the densely woven (and hence brighter) webs more often than the less-dense webs. In the case of pigmented silks, silk color alone was less important than was the ambient light intensity. However, as noted in the experiments below, the colors of the webs might not have been distinguishable by the bees.

Table 6.1 Bee response to webs that differ in color, architecture, or both. (A) Maximum likelihood analysis of variance of rate of interception to avoidance of *T. fuscipennis.* (B) Maximum likelihood estimates for main effects of time, day, site, web and their interactions averaged across all sites.

A.

Source	*df*	χ^2	*P*
Intercept	1	2992	0.00
Minute interval	3	1312	0.00
Day	2	147	0.00
Site	3	139	0.00
Web	2	343	0.00
Minute interval by day	6	11	0.09
Minute interval by site	9	61	0.00
Day by site	6	116	0.00
Minute by interval by web	6	28	0.00
Day by web	4	119	0.00
Likelihood ratio	70	425	0.00

All the main effects were significant. The day by web interaction showed that bees responded differently on different days of the experiment. Although significant, the least important effect was site. Hence, the bees' response to the webs during the different time periods of the experiment was greater than their responses to the feeding stations where webs were present.

B.

Main effects	Estimate	SE
Minute interval		
15	−2.50	0.07
30	−0.52	
60	0.08	
120	2.11	
Day		0.06
2	−0.62	
3	0.01	
4	0.61	
Web		0.06
A. argentata (unpigmented)	−1.20	
N.clavipes (unpigmented)	0.08	
N. clavipes (yellow)	1.16	

Estimates large and positive indicate fewer bees than average are intercepted for condition illustrated. Estimates large negative indicate more bees than average are intercepted. Exact probability calculated by dividing estimate by its SE.

All ratios greater than 1.96 were significant at the 0.05 level.

Taken together, the results of these experiments show that the avoidance response of *T. fulviventris* toward spider webs is not innate to the bee. Most of the insects were intercepted on Day 1 on their first visit to the web, regardless of whether the low- or the high-fiber-density web was used. With both types of web, the daily rate at which insects learn to avoid them depends on the bees' previous experience at the web; that is, the proportion of avoidance increases from the first

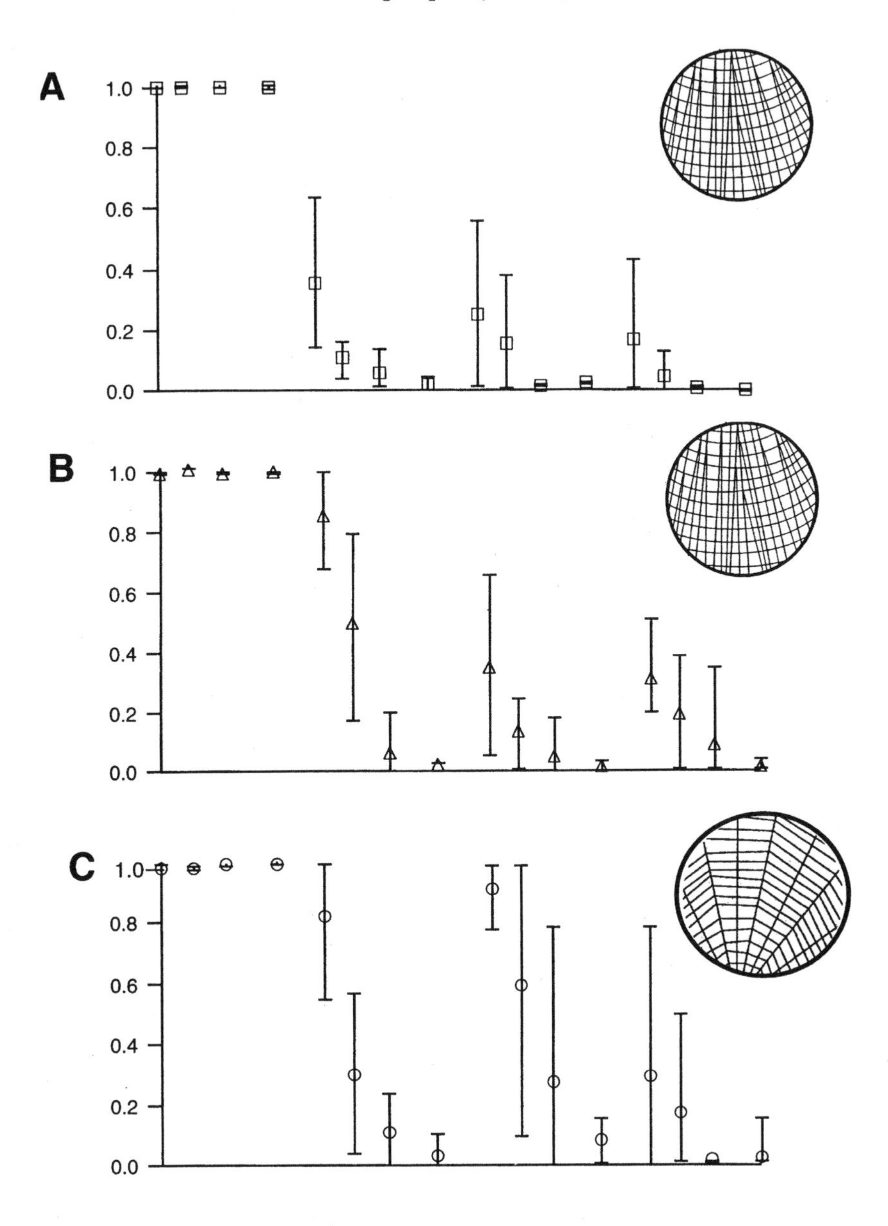

Figure 6.6 Avoidance response of bees to webs that differ in spectral properties (A versus B), in architectural design (B versus C), or both (A versus C). The icon in upper left corner of each graph shows the design of webs spun *Nephila clavipes* (A, B) and those spun by *A. argentata* (C). Web A is yellow; webs B and C are both white. The data collected at the four sites were combined according to web type. For each day, the mean and SD of the proportion of bees flying through the ring or intercepted, retreating or avoiding the web, were calculated from the total number of bees flying through the web. The three horizontal panels compare insect response to web type. The four vertical panels compare bee response by day. The data were collected over a 2-hour period each day and sorted into four intervals of 15, 30, 60, and 120 minutes for analysis. (Redrawn from Craig 1994.)

visit to the second, third, or fourth visit to the site. On the second day, fewer insects were intercepted by the high-fiber-density webs than by the low-fiber-density webs. Fewer bees were intercepted by the yellow webs than by either of the unpigmented high- or low-fiber-density webs. Still, bees took longer to learn to avoid yellow webs than the other two types of web, suggesting that avoidance learning is based on the experiences made during interception. Thus, both fiber density and web color affect bees' response.

It should be noted that the angle at which bees approached the webs was not monitored. Because webs are porous, the viewing angle is expected to affect insect perception. A low-fiber-density web viewed obliquely might be as visible as a higher-density web viewed orthogonally. This effect remains to be examined under controlled laboratory conditions.

Drosophila may be attracted to UV-reflecting silk due to their spontaneous open-space response

Most insect species are innately attracted to emitted UV light (for example, UV lamps), or to surfaces reflecting light that contains UV. Closely related to this phenomenon is the so-called open-space response; that is, the attraction that insects display toward open, bright gaps (Goldsmith 1961; Hu and Stark 1977), usually illuminated by skylight, when flying away from dim, closed sites. Because skylight is rich in UV content, the open-space response might be based on the insect's innate preference for UV. Webs made of UV-reflecting silks, such as those made by members of the superfamily Deinopoidea, might thus be confused with a gap leading to an open space. There are several examples of spiders that weave UV-reflecting webs in the open understory (see chapter 5). The diurnal spider *Hyptiotes* (Uloboridae) makes an orb-web that is reduced to a three-sector triangular net (Opell 1985). *Miagrammopes*, which also is an uloborid, forages in similar forest sites, but at night. It weaves a web that is reduced even further to a single thread or a maze of threads (Lubin 1978). Still other uloborids create artificial gaps by weaving leaf material or detritus into a three-dimensional maze of UV-reflecting silks constructed above an orb catching net (see Lubin 1986). Insects flying away from closed vegetation or from the forest floor may be attracted to the fiber maze and, if intercepted, fall into the sticky orb below it.

To examine the attractiveness of UV-reflecting webs, Craig and Bernard (1990) tested *Drosophila* flies in a series of choice experiments conducted in the laboratory. A light-tight, two-armed Y-maze was built from black Plexiglas. Two identical webs, 12 cm in diameter, built by the same female spider (*Uloborus glomosus*) were presented on Plexiglas rings to the spiders, one in each of the arms.

To eliminate the possible effect of pheromone attractants, the webs were isolated from the flies by a piece of UV-transmitting Plexiglas. One web was illuminated by white light from which all wavelengths shorter than 405 nm were filtered out. Hence, the web reflected light between 405 and 650 nm. The other web was illuminated by white light that passed through a UV-transmitting filter and thus contained all wavelengths between 300 and 650 nm.

Several investigators have shown that the open-space response need not be correlated with UV-reflectance, but might well be based on positive phototaxis; that is, on the animals' tendency to move from dim to bright sites (Labhart 1974; Rose and Menzel 1981; Menzel and Greggers 1985). Because the beam that did not contain UV was dimmer than the one that contained all wavelengths, the latter might be more attractive simply because it was brighter. To account for this, the light intensity of the two light beams was measured and equalized. Thus, the two webs differed only in the UV content, but not in brightness.

Drosophila were released into the unlit base of the Y-maze and flew immediately in the direction of the light. At the bifurcation point, both webs could be seen simultaneously. In all ten trials conducted, the flies preferred the arm that contained the UV-reflecting web. This experiment was extended by Watanabe (1999) who compared the effects of web decorations spun by the uloborid spider *Octonoba sybotides* on the insect's response. Web decorations are woven from UV-bright silks and are much denser than the rest of the web surface. In this experiment, the flies preferred the decorated webs that reflected more UV-light than the alternative webs that lacked decorations.

It may be that UV-reflecting webs only attract insects over a range that is too long for the web pattern to be resolved. At a close range, insects might be able to recognize the web pattern and flee. To control for this possibility, flies were introduced into a maze in which the Plexiglas barriers that separated them from the webs (see above) were removed, thus allowing the flies to come very near to the web. When the web was illuminated with light that did not contain UV (and thus did not reflect UV), insects turned away from it and escaped capture. However, when the same web was illuminated with light that contained UV, the flies flew into it and were intercepted (Craig and Bernard 1990).

In the choice experiments described above, the only factor affecting insect response was whether or not the webs reflected UV light. Many small insects, especially those caught in webs spun by uloborids, live and forage in leaf litter on the forest ground. When these insects are flying from site to site, they must fly upward, toward open space. Whenever an uloborid web is illuminated by daylight, it represents a UV-bright site, signaling open space. Hence, sunflecks in which UV silks are suspended are likely to attract insects. The attraction exerted by these silks might have been part of the particular circumstances that allowed the Deinpoidea, the ancestral orb-spinners, as well as other cribellate spiders, to persist through time despite the evolution of the apparently more efficient webs spun by the araneoeids (see chapter 3). One could speculate that the Uloboridae have specialized on insects that are in the open-space response mode. They may optimize their foraging success in sites where the web surround is dim, yet the web itself is illuminated. Sunflecks are likely to be abundant in forests, otherwise no plants would be found in the understory. The probability that a sunfleck will strike the surface of a web depends on the angle of the sun, on the density of the leaf overstory, and on the frequency with which the wind perturbs it.

Studies on web avoidance learning show that bees are able to dissociate color cues from the information with which it is paired

Most studies on visual performance of honeybees have exploited this insect's excellent learning capacity. Floral color signals are well represented in the bee's perceptual color space (see figure 6.3). The capacity to form food-color associations very rapidly is pre-programmed and allows bees to forage successfully with little previous experience (Menzel 1985). Furthermore, honeybees are innately predisposed to learn to associate some colors (for example, UV-violet and blue-green) with nectar rewards faster than others (Menzel 1967; Menzel et al. 1974). Nonhymenopteran pollinators, such as flies and beetles, prefer blue-green and yellow flowers (Chittka 1996). White flowers are visited by all groups of insects. In addition to color, bees show a predisposition to learn particular shapes, mainly symmetrical ones (Lehrer et al. 1995; Giurfa et al. 1996a).

If bees use the same cues to associate flowers with a food reward and to associate webs with punishment, then they are expected to learn rapidly to avoid spider webs whose colors or shape are similar to flower colors and shapes. This is not to say, however, that webs are confused with flowers, although web decorations may be (see below). Rather, bees might be able to dissociate the type of information; that is, positive or negative, from the color or pattern signal with which it is paired. When spiders fail to capture insects that are intercepted by webs, the insects have a chance to associate the color or pattern of the web with danger and learn to avoid similarly colored webs and patterns on future encounters. This association, however, should not result in the avoidance of flowers that possess a similar color or pattern.

Because bees are color-sensitive (Chittka and Menzel 1992), spiders that spin pigmented webs may attract prey or use the color to camouflage the web against background vegetation. An evolutionary cost of colored silks is incurred, however, if colored silks help insects learn to avoid webs. From the spider's perspective, this cost is balanced if the total number of insect interceptions per colored web is higher than the number of interceptions that would occur if the web were not colored. In this case, it does not matter whether the insect was intercepted on its first, second, or third encounter with the web. From the insect's perspective, on the other hand, avoidance should be learned in a single trial; that is, the insect must learn the color (or the shape) of the web on its very first interception with the web. Because the behavior of both spiders and insects affects the evolution of silk color, data on insect learning and interception need to be analyzed from both the spider's and the insect's perspective.

Such an analysis was conducted by Craig et al. (1996), who investigated the interactions between color, avoidance, and reward by comparing the bees' response to artificially colored webs (figure 6.7). All bees in the experiment were marked to be recognizable individually, and were trained to sites where they were presented with violet, blue, yellow, green, red (bee-black), and bee-white webs.

Comparison of the total numbers of bees responding to the colored webs (table 6.2a) showed that bees were intercepted more often at yellow webs than at webs of the other colors (red, blue, blue-violet, and white). Analysis of the bees' learning rates shows that they learned most rapidly to bypass blue webs (table 6.2b). Comparison of the subset of data that includes bees' behavior after their first interception (web effect viewed from the insect's perspective) showed that they rapidly learned to avoid violet and red (bee-black) webs. There is no evidence that they learned to bypass or retreat from white, blue, and yellow webs (Craig et al. 1996).

It should be noted that, in these experiments, the webs were not balanced for total reflected light. It is known, however, that intensity plays only a minor role, if any, when color learning is involved (see chapter 5).

The finding that bees showed only a poor learning capacity for the yellow webs was explored further by making use of the naturally pigmented webs spun by *N. clavipes* (described in figure 6.1). Bees were trained to forage at forest and nonforest sites and then presented with pigmented and unpigmented webs. In the four different experimental set-ups, bees viewed colored webs at sites where they were normally found, and at sites where they would not be found in natural conditions.

The results of these experiments (figure 6.8) show that in dim light (where free-ranging spiders produce only unpigmented silks), regardless of whether the webs were pigmented or not, the proportion of intercepted insects was similar to that of insects that learned to avoid the webs (figure 6.8A) (Craig et al. 1996). In bright light, however (figure 6.8B), bees responded to the two types of web differentially. Although almost all insects were intercepted the first time

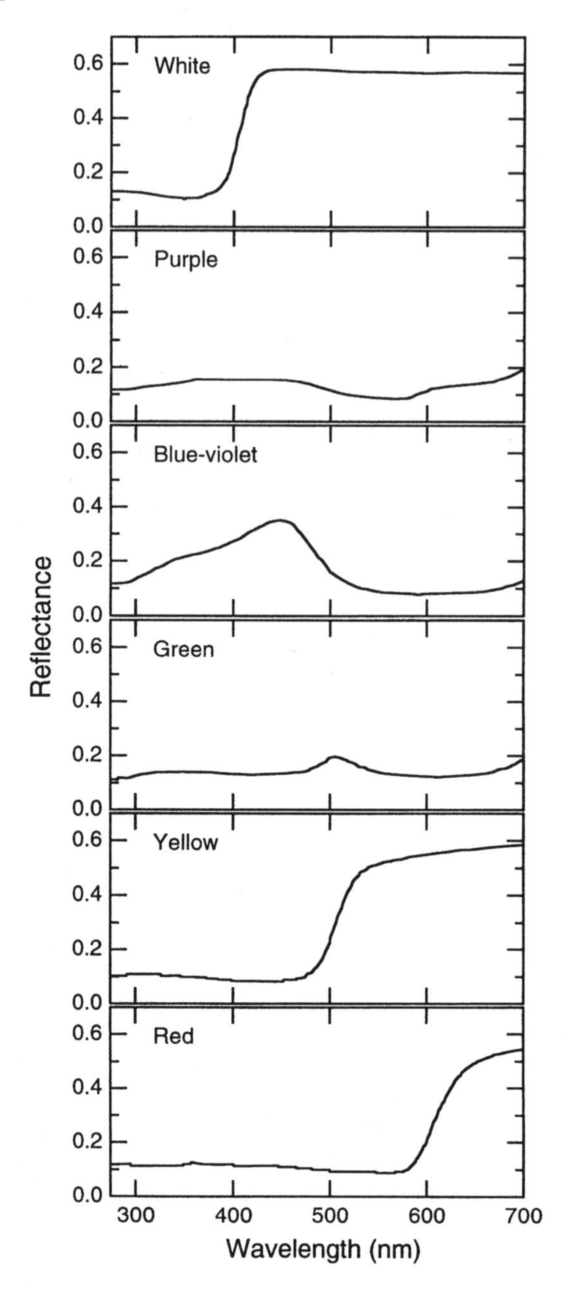

Figure 6.7 Spectral properties of paints used to color webs. Powdered tempera paint and confectioners' sugar were aspirated onto webs, coloring them blue, violet, yellow, green, and red. While the pigments altered the spectral properties of the webs, the confectioners' sugar hydrated, allowing webs to maintain their viscid properties. In addition, titanium oxide (rutile) was aspirated onto webs to enhance their total reflectance while maintaining the natural spectral properties of the silk. These webs appeared bee-white. (From Craig et al. 1996.)

they encountered an unpigmented web, in all subsequent encounters they avoided them. In contrast, only about 20% of the bees were intercepted the first time they encountered yellow webs. These results suggest that pigmented webs are more visible to the bees than are nonpigmented webs. Interestingly, however, the bees learned to avoid the yellow webs significantly more slowly than they learned to avoid the unpigmented webs (figure 6.8A and B).

These results suggest that bees learn to pair color cues (for pattern, see below) with both negative and positive stimuli, and that their learning performance is comparably fast in both cases. Thus, honeybees are attracted to and learn rapidly to feed from blue-violet stimuli, and stingless bees learn rapidly to avoid blue-violet spider webs. Honeybees learn more slowly to feed at yellow stimuli, and stingless bees learn more slowly to avoid yellow spider webs (table 6.3).

It is possible that, during evolution, the speed at which learning of a color occurs has been influenced by the predictability of color-object associations. For example, the probability that violet is associated with any naturally occurring object other than a flower is low. A long-wavelength color such as yellow, on the other hand, is more ambiguous owing to its occurrence in many terrestrial objects, and not only in flowers. Depending on the web background, spiders may have evolved to produce pigments that attract

Table 6.2a Frequency with which stingless bees are intercepted at artificially colored webs. (A) From the spider's viewpoint (that is, regardless of the number of times the insect viewed the web prior to interception). (B) Bee learning. (From Craig et al. 1996.)

Effect	Estimate	Error	χ^2	*P*
Intercept	2.7	.15	340.0	<.0001
Day 1	−.84	.13	42.7	<.0001
Day 2	.84	.13	42.7	<.0001
Visit 1	−1.12	.16	46.1	<.0001
Visit 2	−.07	.20	.10	.7483
Visit 3	−.10	.20	.23	.6298
Visit 4	.19	.22	.73	.3932
Blue violet	−.32	.23	2.0	.1608
Blue	−.30	.20	2.3	.1292
Green	−.07	.28	.05	.8156
Red	.56	.21	6.7	.0097
White	.98	.27	13.0	.0003
Yellow	−.85	.19	20.8	<.0001

Note: Negative numbers indicate that more bees were intercepted at the webs than bypassed them. Significantly more bees are intercepted at yellow webs, and significantly fewer bees are intercepted at red and white webs. The number of bees intercepted by blue violet, blue, and green webs does not differ from random.

Table 6.2b Ability of insects to learn to avoid artificially colored webs. From the spider's viewpoint, it doesn't matter when during a foraging period the insect is intercepted, simply that is intercepted. Therefore, from the spider's perspective and using all approach data, we calculate that bees rapidly learn to bypass blue violet and blue webs but learn only slowly to bypass yellow, red and white webs. From the insect's viewpoint, however, learning is calculated post-stimulus, i.e. subsequent interception probability after the insect's first interception. Bees learn rapidly to bypass blue violet and red webs, but learn only slowly to bypass green webs. The probability that bees will learn to bypass yellow, blue and white webs does not differ from random.

Effect	Estimate	Error	χ^2	*P*
Spider's viewpoint:				
Intercept	−1.9	.11	293.6	<.0001
Day 1	.45	.09	24.7	<.0001
Day 2	−.45	.09	24.7	<.0001
Visit 1	.84	.13	42.1	<.0001
Visit 2	.26	.14	3.4	.0673
Visit 3	−.28	.16	3.1	.0799
Visit 4	−.81	.19	18.4	<.0001
Blue violet	.89	.16	29.4	<.0001
Blue	.99	.16	39.7	<.0001
Green	−.52	.30	3.1	.0804
Red	−.72	.20	12.8	.0003
White	−.50	.20	6.2	.0132
Yellow	−.86	.19	20.8	<.0001
Bee's viewpoint:				
Intercept	−2.56	.18	208.0	<.0001
Day 1	.42	.12	11.9	.0006
Day 2	−.42	.12	11.9	.0006
Visit 1	1.06	.17	37.7	<.0001
Visit 2	−.12	.22	.30	.5818
Visit 3	−.44	.23	3.6	.0593
Visit 4	−.51	.24	4.6	.0316
Blue violet	.86	.24	12.5	.0009
Blue	−.46	.35	1.8	.1858
Green	−1.60	.61	6.9	.0088
Red	1.20	.24	24.2	<.0001
White	−.06	.26	.06	.8000
Yellow	.07	.29	.06	.8051

prey to their webs (when web and background contrast are high) as well as to make their webs difficult to learn to avoid. Indeed, a new way to look at spiders would be to classify their foraging behavior not in terms of the different taxa of prey that they catch, but rather by the properties of the insects' visual systems, their learning capacities and foraging ecologies.

Web decorations attract prey and their variable orientations may disrupt insect pattern learning

Bees discriminate well among different shapes and patterns (see, for example, Campan and Lehrer 2002; for reviews of older literature, see Wehner 1981). Furthermore, bees have been shown to spontaneously prefer some shapes over others (Hertz 1930; Anderson 1997; Lehrer et al. 1995; Giurfa et al. 1996). Similar to color learning, some patterns are learned faster than others. Although insects tend to rely on color signals more strongly than on spatial parameters, the latter often provide the only cues that insects use for orientation.

Many spiders decorate their webs with bright silk patterns (see above) that produce contrast against their background and seem to attract prey. In some cases, these decorations resemble the types of floral patterns that were shown to attract bees attract bees (Craig and Bernard 1990; Lehrer et al. 1995; Giurfa et al. 1996). As in the case of color learning, however, if an intercepted insect escapes capture, then the decorative patterns are memorized as aversive signals, and the insect will avoid them in the future (figure 6.9).

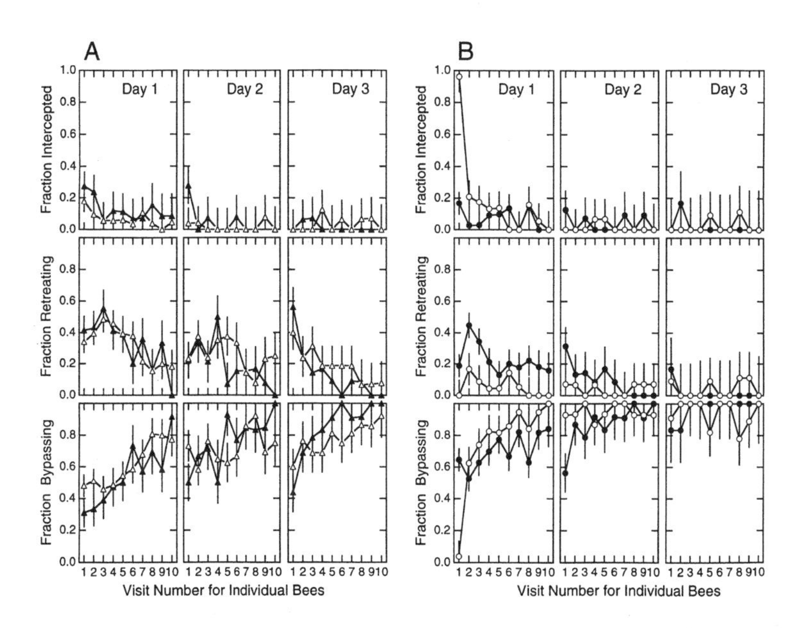

Figure 6.8 Quantification of behavior of stingless bees at natural, yellow pigmented and unpigmented webs (mean values and SDs based on the behavior of individually observed bees). (A) Total number of bees intercepted at webs as a function of the number of visits made to the experimental feeding stations. From the spiders' point of view, it does not matter whether interception occurs on the insect's very first arrival, or on a later one.

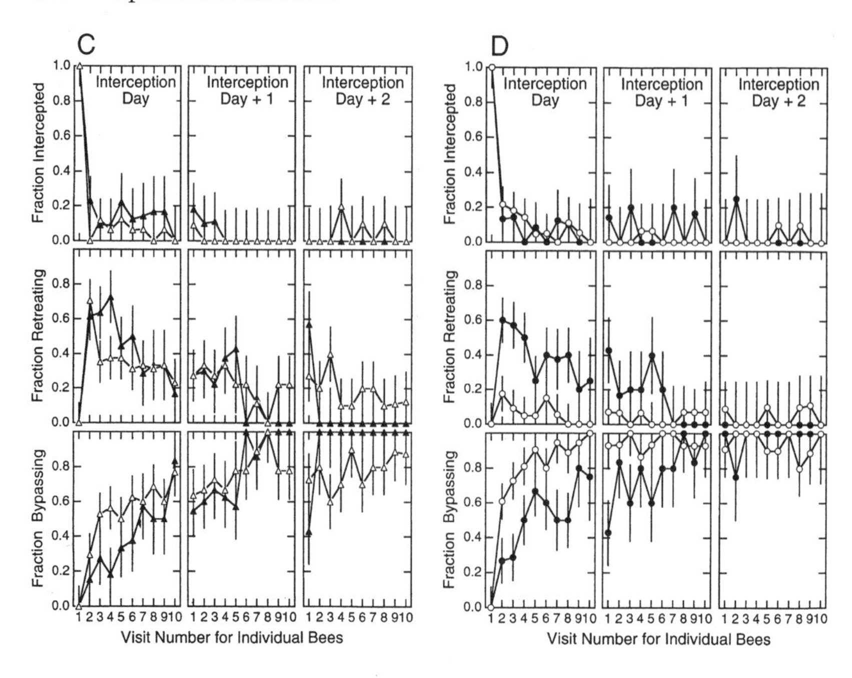

Figure 6.8 (*continued*) These graphs show that, in dim sites (A), bees respond similarly to yellow pigmented (filled triangles) and unpigmented webs (open triangles). (See Table 6.2.) (B) Same as in (A), but experiments were conducted at open, bright sites. On their first arrival, a much larger proportion of bees intercepted at unpigmented webs (open circles) than at yellow webs (black circles), indicating that pigmented webs are easier to see. (C) Post-interception behavior of bees at unpigmented and yellow webs. From the bees' viewpoint, it is important to learn to avoid the web after one interception. When foraging in dim light, even after having been intercepted once, bee have more difficulty learning to avoid yellow webs (filled triangles) than unpigmented webs (open triangles). (D) When foraging in bright light, bees have even more difficulty learning to avoid yellow webs (filled circles) than they did in the forest understory. Unpigmented webs are bypassed to reach the dish (open circles). (From Craig et al. 1996.)

Table 6.3 Summary of bee response to *Nephila* yellow pigmented and unpigmented webs in bright and dim light. (From Craig et al. 1996.)

	Spider's viewpoint		Bee's viewpoint	
Web condition	Dim	Bright	Dim	Bright
Unpigmented	*Repeatedly approached*	Bypassed	*Slow avoidance learning*	Rapid avoidance learning
Pigmented	Repeatedly approached	*Repeatedly approached*	Slow avoidance learning	*Slow avoidance learning*

Note: Italic entries indicate natural web color in usual light conditions. For example, spiders normally spin pigmented webs in bright light and stingless bees normally encounter pigmented webs in bright light.

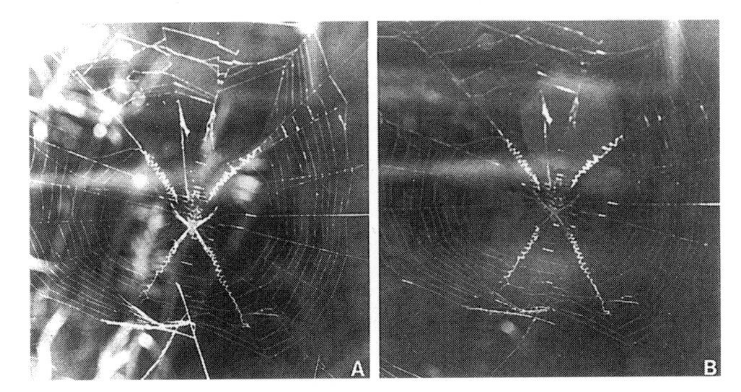

Figure 6.9 Web decorations photographed in visible light (A) or in UV light (B). Background vegetation that appears bright in visible light is dark or disappears completely in the UV [note grass flowerhead and stem in (A), and absence of stem in (B)]. A plot of spectral data for decoration silk and viscid silk reveal high contrast in the UV. (From Craig and Bernard 1990; reprinted with permission from The Ecological Society of America.)

Web decorating is a common foraging strategy among spiders. In fact, seventy-one species of orb spinner drawn from nineteen different genera decorate their webs. A phylogenetic analysis suggests that web-decorating behavior has evolved independently nine times in the Orbiculareae, and has been lost once (figure 6.10) (Scharff and Coddington 1997; Herberstein et al. 2000a,b). Spiders in the genus *Argiope* decorate their webs with UV-reflecting silks spun into patterns such as spirals, crosses, tufts, and dabs (figure 6.11). These decorative patterns resemble the UV light-and-dark "nectar guides" found on flowers that have evolved to signal the location of nectar and pollen.

We showed that insects are attracted to web decorations spun by *A. argentata* (Craig and Bernard 1990), by using either natural or experimentally manipulated decoration patterns to compare insect interception rates between decorated and undecorated webs. Because the web site greatly affects the spider's foraging success, the experiments were designed to control for site effects by providing a within-site control. At least one-half of each web was undecorated. The number of insects intercepted by decorated web halves was then compared with the number of insects intercepted by undecorated web halves. The results showed that, after correcting the data for variation due to web site alone, significantly more insects were intercepted at decorated webs than at undecorated webs. Specifically, 1.27 insects per web per hour were intercepted at undecorated web halves, and 2.54 insects per web per hour at decorated web halves. Adding a spider to the decorated webs resulted in an average of 3.26 insects per web per hour, or 58% more insects than were intercepted at undecorated webs without spiders.

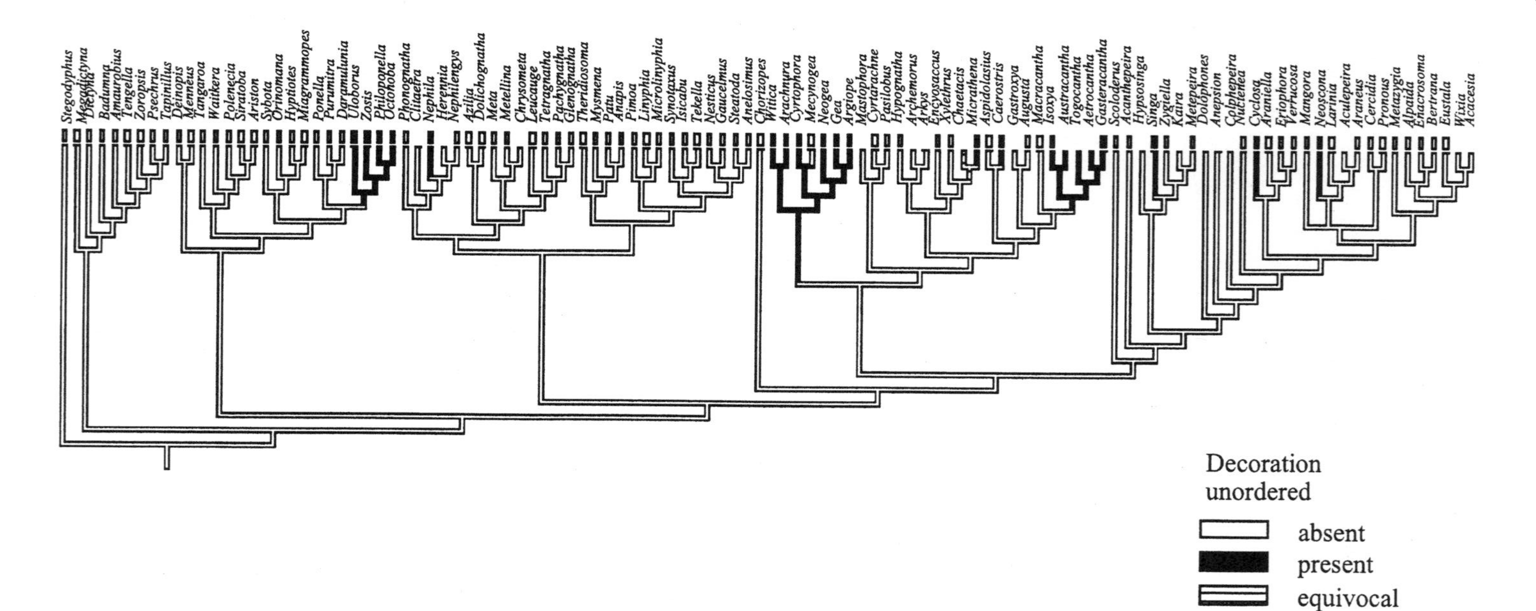

Figure 6.10 Character optimization for web decorations on equally weighted tree. Tree determined by maximum parsimony; no additional information regarding branch support available, but see Scharff & Coddington, 1997. (From Herberstein et al. 2000, with permission from Cambridge University Press.)

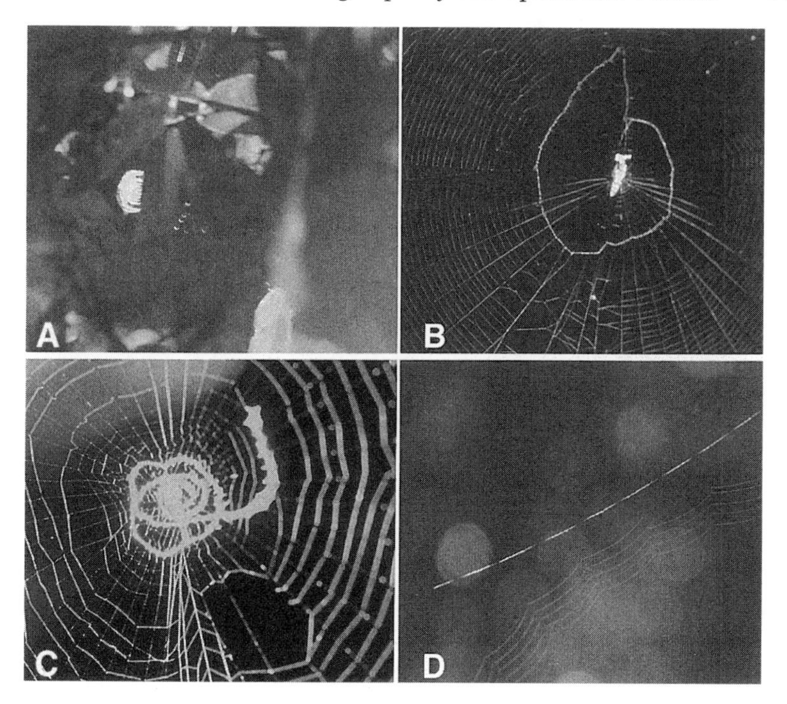

Figure 6.11 Photographs of several natural web decorations. (A) Decoration ~4 meters (12 feet) over above the forest floor; (B) decoration by *Cyclosa*; (C) decoration by *Philoponella tingens*; (D) decoration by *Witica*.

In the light of the enhanced success of foraging with the help of decorated webs, it would be expected that *A. argentata* might decorate their webs daily, but they do not. The population of spiders we studied on Barro Colorado Island was highly polymorphic for decorating behavior (figure 6.12). *A. argentata* show individual variations in the propensity to decorate, suggesting that web decorating has a heritable component (figure 6.13; table 6.4). Some lineages of spiders rarely decorate, some decorate quite frequently, whereas others decorate at intermediate intervals. Individual spiders also vary the orientation and patterns of their decorations. The obvious strategy to maximize prey capture, decorate a web daily, has not evolved in this species. Thus, if insects keep returning to the same foraging site day after day, as do bees, then they are likely to find either no decoration or a different pattern than the one they have previously learned to avoid.

Although the tendency to decorate is inherited, it seems that environmental factors also play a role. We maintained spiders in the laboratory under controlled feeding regimens and found that the decoration patterns differed in response to diet. Spiders on a high-prey diet decorated 20% of all new webs, whereas spiders

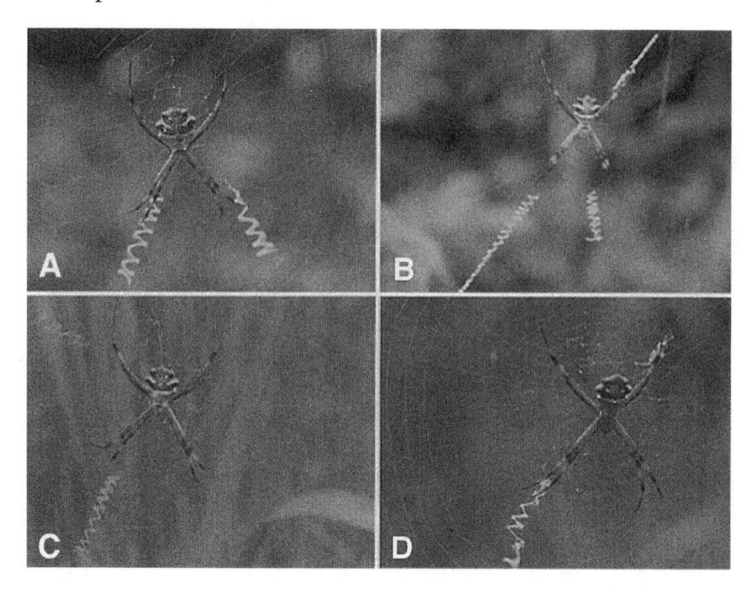

Figure 6.12 A sample of patterns of web decorations spun by spiders *Argiope argentata*. Among adult *A. argentata* found on Barro Colorado Island, sixteen different types of web decorations composed of one to four arms of a cross were found. In most cases, the arms were oriented in the clock positions 2, 5, 7, and 10. (Reprinted with permission from *Behavioral Ecology and Sociobiology*, Volume 35 © 1994, Springer-Verlag. "Predator foraging behavior in response to perception and learning by its prey: interactions between orb-spinning spiders and stingless bees." Pp. 45–53. Figure 1.)

maintained on a low-prey diet decorated only 11% of all new webs (ANOVA for whole model $n = 35$, $r^2 = 0.74$, $F = 144$, $P<0.0001$; effect of food regimes, Type 1 SS, $F = 529$, $P<0.0001$). Herberstein et al. (2000) found a similar response pattern for the closely related spider *A. keyserlingi*, but they also demonstrated that spider behavior varied depending on whether prey were offered on a predictable or random schedule. Both of these studies showed that decorating behavior can be induced in spiders that are maintained on a rich diet. In addition, spiders maintained on a high-prey diet spun significantly smaller webs even though they were more likely to decorate them (Hauber 1998; Craig et al. 2001). At first sight, this finding seems to be contrary to the expectation, but it is actually not. When spiders are hungry, large webs are built to intercept any insect that might enter the web area. When spiders are less hungry, on the other hand, they make smaller webs that are decorated to attract particular types of prey (Craig et al. 2001).

To quantify the effect, if any, of pattern learning on the rate of insect interception, we designed a field experiment in which webs in a 100 × 50 m field were divided in two populations. In one population, all webs were decorated identically for two days in a row. In the second population, all decorations were oriented randomly for

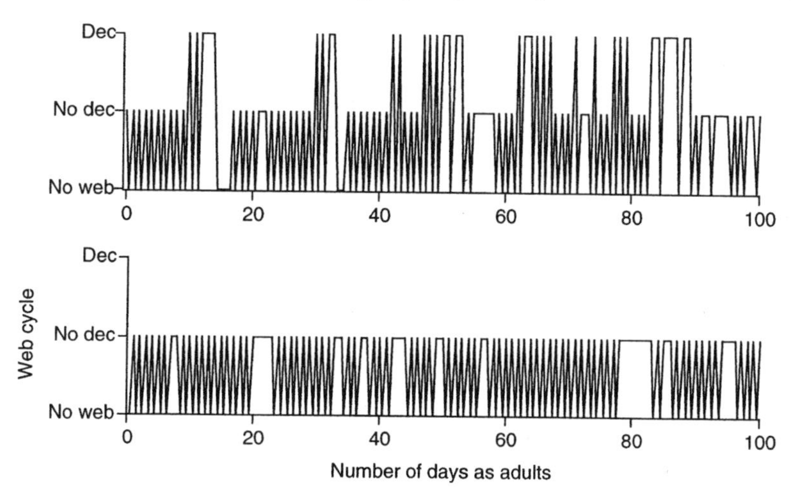

Figure 6.13 Individual spiders display variable decorating behavior. The decorating activity of a laboratory population of adult *Argiope argentata* that were maintained on a constant diet of crickets. The data illustrate the web spinning and decorating behavior of two spiders over a 100-day period. Spiders either wove new webs but did not decorate them (a spike to the "No dec" position on the Y-axis), they removed their webs (trace returns to "No web" position), or they retained the web for two or more days (trace lies parallel to the X-axis), or they wove a decorated web (spike to "Dec" position on the Y-axis). (From Craig et al. 2001; reprinted with permission from *Evolution,* Allen Press.)

two days in a row. When insect interception rates were compared, the rate of prey interception on Day 1 was the same in both populations. On Day 2, however, webs that were decorated as they had been on the previous day showed significantly lower rates of insect interception than did webs whose decoration orientations had been changed. The variables of web site, day of experiment, and the number of visits are all significant and of approximately equal importance to the probability of bee interception. The effect of the decoration orientation, either varying or constant, however, is by far more significant than any of the other variables (Craig 1994b).

To explore these results in more detail, we extended this work to test the behavior of one of the most important prey of *A. argentata*, the stingless bee *Trigona fuscipennis.* The second experiment was based on a training procedure with individually marked stingless bees. We trained the bees to forage at two sites where they were confronted with decorated webs (figure 6.14). As described above, the bees were trained to feed at an artificial feeding station by flying through an open, black ring (see figure 6.5). During the control trials, the bees encountered webs with decorations that changed their orientation every day. During the experimental trials, the decoration orientation was kept constant.

The results showed that orientation of the web's decorations affects the rate at which individual bees return to the web site, as well as the rate at which they are intercepted. When the effects of

Table 6.4 Pedigree data for web decorating behavior of adult, female offspring. Males are classified by the decorating behavior of known mothers. All females mated with just one male, although multiple offspring may originate from one or more egg sacs. The number of decorated webs spun by of each of the offspring was normalized and the spiders were divided into three groups: spiders that decorated their webs rarely or not at all (<25% of all new webs decorated, L); spiders that decorated their webs with moderate frequency (25–75% of all new webs decorated, M); and spiders that decorated their webs almost every day (>75–100% of all new webs, H). (From Craig et al. 2001.)

Female ID	Female decorating frequency (%)*	Male decorating frequency (%)*	Offspring decorating frequency
Females that decorate at high-frequency x with males whose mothers decorate at medium frequency			
AB98	96	45	H
Females that decorate at high-frequency x with males whose mothers decorate at low ferquency			
AR50	96	4	L,M,M,M
AB22	91	0	L,M,H
AG67	89	5	L
AG37	81	5	L
AB1	80	5	L,L,L
Females that decorate at medium-frequency x with males whose mothers decorate at high frequency			
AG2	65	89	M,L
Females that decorate at medium-frequency x with males whose mothers decorate at low frequency			
AG99	52	25	L,L,L,L,M
AW83	65	6	L,L,L,L,L,L
Females that decorate at low-frequency x with males whose mothers decorated at high frequency			
AY91	0	89	L,L,H
Females that decorate at low-frequency x with males whose mothers decorated at medium frequency			
AY44	0	52	M
Females that decorate at low-frequency x with males whose mothers decorated at low frequency			
AW76	0	0	L,L,M,M
AR59	0	0	L,L,L
AG87	0	0	L,L,L
AY36	24	25	M,L,L
AY77	19	0	L,L,L,L,L,L
AR56	19	0	L
AG46	0	0	L,L

* Both females and males were mated only once.

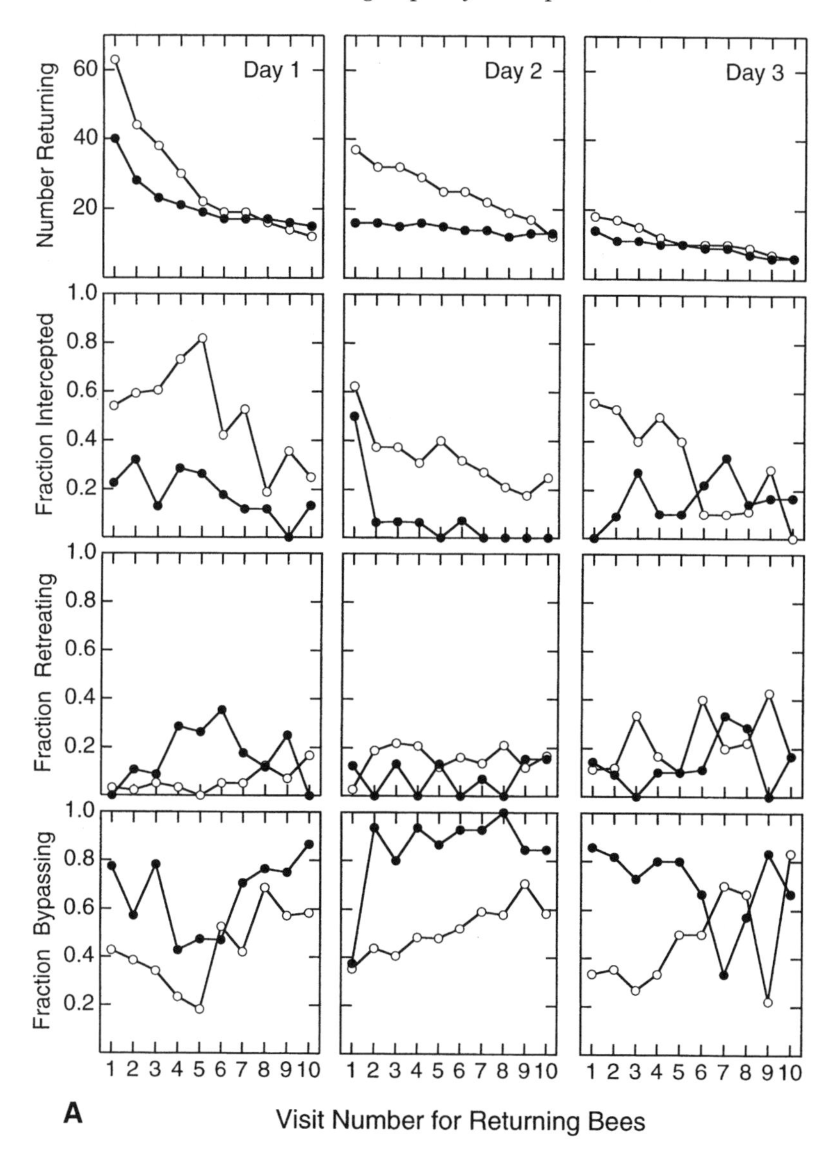

Figure 6.14 Bees rapidly learned to avoid webs that were decorated the same way each day (filled symbols), but had difficulty learning to avoid webs whose decoration orientations varied (empty symbols). The data are plotted in two ways, (A) and (B). The sequences in (A) illustrate the effect of decorated webs on all responses of bees. The sequences in (B) show the effect on the bees' response after they had been intercepted once. (Reprinted with permission from *Behavioral Ecology and Sociobiology*, Volume 35 © 1994, Springer-Verlag. "Predator foraging behavior in response to perception and learning by its prey: interactions between orb-spinning spiders and stingless bees." Pp. 45–53. Fig. 3A,B.)

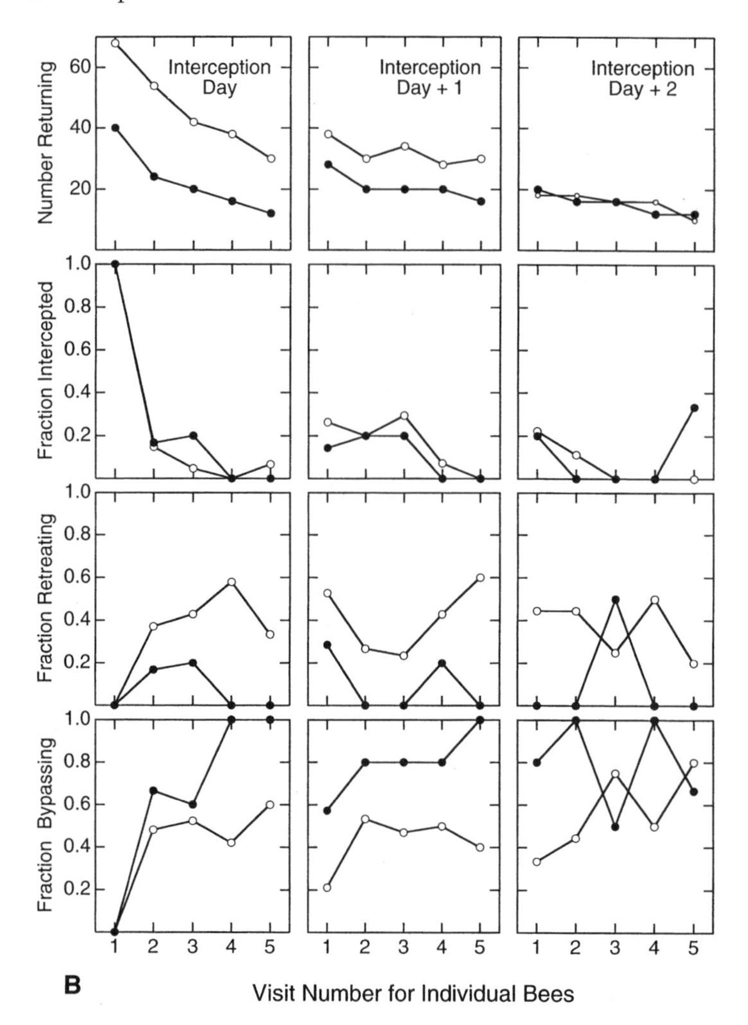

Figure 6.14 (continued)

web site, day, the number of times bees were observed at the web, and web decoration effects were compared, the effect of the decoration orientation, either varying or constant, proved again to be the most important variable affecting bee behavior (table 6.5a). Furthermore, maximum likelihood estimates show that significantly more bees are intercepted during the first to the fourth visit to the webs, and that significantly fewer bees are intercepted during the seventh to tenth visits. A post-interception analysis (table 6.5b) shows that most bees learn to bypass webs after only one interception. These data suggest that avoidance learning at spider webs may be as fast as reward learning (Menzel 1967; Menzel et al. 1974).

In a subsequent experiment with different populations of bees, we tested to see if bees could generalize a once-learned avoidance behavior at a particular training site to decorated webs presented at a different site. We trained bees to forage, sequentially, at four different sites at four different times in the morning. The webs presented at the four sites were decorated identically. We found that the bees were not able to generalize learned avoidance behaviors, even when sites were closely grouped. Thus, the aversive stimulus is stored in memory together with the site at which it has been learned, a finding already reported for learning by reward (see, for example, Collett and Kelber 1988).

Table 6.5a Statistical analysis of the effect of orientation of web decorations on the behavior of all bees (from the point of view of the predator). The variables web site, day of experiment and bee visit number are significant and of approximately equal importance to the probability of bee interception. The effect of the decoration orientation, either varying or unchanging, is much more important than any of the other variables. The maximum likelihood estimates show that bees learn to avoid the web by their fourth visit and that significantly more bees are intercepted when the orientation of the web decoration changes from day to day. (From Craig 1994.)

A. Maximum likelihood analysis of variance

Source	*df*	χ^2	*P*
Intercept	1	129.34	<0.001
Site	1	27.20	<0.001
Day	2	23.70	<0.001
Visit number[a]	4	31.91	<0.001
Decoration effect	1	94.44	<0.001
Likelihood ratio	51	103.66	<0.001

B. Analysis of maximum likelihood estimates

Effect		Estimate	SE	χ^2	*P*
Intercept		1.07	0.09	129.34	<0.001
Site	1	−0.40	0.08	27.20	<0.001
	2	0.40	0.08	7.20	<0.001
Day	1	−0.48	0.10	22.72	<0.001
	2	0.12	0.11	1.25	0.26
	3	0.37	0.11	10.69	<0.001
Visit[a]	1	−0.54	0.13	17.49	<0.001
	2	−0.49	0.14	11.76	<0.001
	3	−0.11	0.16	0.44	0.50
	4	0.37	0.18	4.21	0.04
	5	0.77	0.23	11.50	<0.001
Changing decoration		−0.82	0.08	94.44	<0.001
Constant decoration		0.82	0.08	94.22	<0.001

[a] Visit numbers 1&2, 3&4, 5&6, 7&8, 9&10 grouped for analysis.

Table 6.5b Effect of orientation of web decoration on behavior of intercepted bees (from the point of view of the prey). Although invariant or varying decoration orientations have a significant effect on bee response to the web, after the bees have been intercepted once the effect of visit number and the day on which the behavior occurs are stronger. The maximum likelihood estimates show that bees have learned to avoid the web by their fourth visit and that significantly more bees are intercepted at webs when the orientation of the decoration changes from day to day.

A. Maximum likelihood analysis of variance

Source	*df*	χ^2	*P*
Intercept	1	80.00	< 0.001
Site	1	1.74	0.19
Day	2	35.09	< 0.001
Visit number[a]	4	136.91	< 0.001
Decoration effect	1	6.69	0.01
Likelihood ratio	36	139.48	< 0.001

B. Analysis of maximum likelihood estimates

Effect		Estimate	SE	χ^2	*P*
Intercept		2.24	0.25	80.00	< 0.001
Site	1	0.25	0.12	1.74	0.19
	2	−0.25	0.12	1.74	0.19
Day	1	−1.04	0.18	34.40	< 0.001
	2	−0.03	0.19	0.02	0.88
	3	1.07	0.23	18.53	< 0.001
Visit[a]	1	−2.79	0.25	127.66	< 0.001
	2	−0.13	0.26	0.23	0.63
	3	−0.11	0.28	0.16	0.69
	4	2.11	0.59	12.83	< 0.001
	5	0.91	0.38	5.61	0.02
Changing decoration		−0.35	0.14	6.69	0.01
Constant decoration		0.35	0.14	6.69	0.01

[a] Visit numbers 1&2, 3&4, 5&6, 7&8, 9&10 grouped for analysis.

Decorative silk patterns that attract prey are also likely to attract the predators of spiders

Avoidance learning by insects (see above) may not be the only disadvantage resulting for spiders that decorate their webs. Another disadvantage might arise if the spider's predators are attracted to the web in addition to prey (Craig et al. 2001). The only observed predators of *A. argentata* are wasps (and at least once, a coati). Specifically, these include wasps in the families Sphecidae and Vespidae (Blackledge and Pickett 2000). Preying mantids and jumping spiders, which feed on other species of *Argiope*, are also attracted by web decorations (Bruce et al. 2001; Sheah and Li 2001).

The attraction that decorated webs exert on the spiders' predators might act to shorten the spiders' life span. If it indeed does, than the fact that decorating spiders persist, suggests there is some evolutionary advantage to high decoration frequency. It may be that the shortened life of decorating spiders is compensated by faster maturation and earlier reproduction. Faster growth rates are correlated with the rate of prey capture and hence web decorations. In other words, the spider's decorating activity is expected to be correlated with the abundance of prey. This is indeed the case.

To test for a correlation between decorating frequency and prey presence, we monitored the decorating behavior of individual spiders that were relatively undisturbed during the 1993 and the first half of the 1994 field seasons. During the second half of the 1994 season, however, we manipulated the presence of bee populations by cutting the field completely. Subsequently, spiders were tracked to determine if their decorating behavior correlated with the regeneration and flowering of grasses in the field and the fluctuating abundance of stingless bees (figure 6.15).

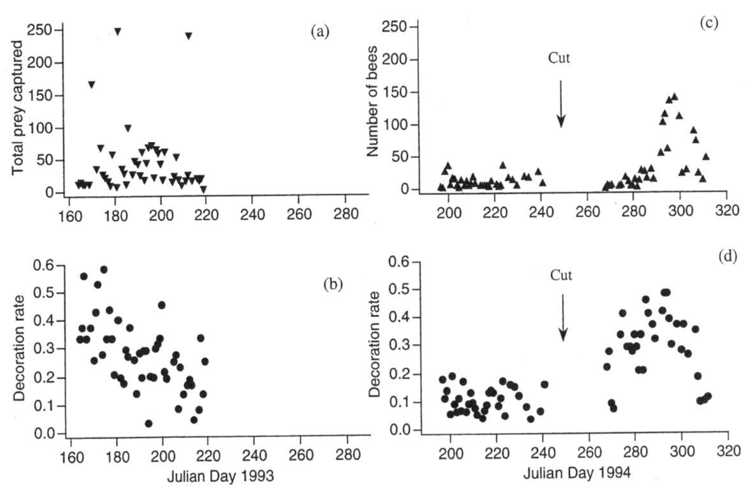

Figure 6.15 The relationship between decorating activity and abundance of prey. (A,B) Comparison of the proportion of decorated webs per day and the total number of bees captured per day (in 1993). There is no clear relationship between web decorating and the number of bees collected from the webs. (C, D) In 1994, during the period before the grass in the field was cut, the decorating behavior of spiders and the abundance of bees varied, at best, inversely. This may be due to site effects, that is, variation in the occurrence of grasses at different microhabitats within the field. During the period after the field was cut, on the other hand, as the grasses recovered and began to flower homogeneously, the decorating activity of the spiders increased with the increase in the abundance of bees. The number of bees sampled in the field each day and the proportion of spiders that decorated webs now reveal a positive correlation. (From Craig et al. 2001; reprinted with permission from *Evolution*, Allen Press.)

An ANOVA of data gathered during the pre-cut and post-cut periods of the 1994 field season showed that the decorating behavior of the spiders differed significantly between the two periods (whole model, $r^2 = 0.6$, $n = 65$, $P<0.0001$; effects, Type 1 SS, $P<0.0001$). In fact, during the first half of the season (the field had already flowered before the study began), decorating behavior and the presence of bees appear to vary inversely (see figure 6.3A,B). After the field was cut and during its regeneration, however, decorating behavior correlated with the flowering of the grasses when the presence of bees in the field reached a maximum (whole model, $r = 0.34$; effects, Type 1 SS, $P<0.02$; figure 6.15).

At the same time, a high decorating activity is expected to increase the probability that an individual spider would be sighted by predators. The data reveal that spiders decorating 75% or more of their webs suffer significantly reduced survivorship (figure 6.16). Spiders that decorated their webs at the lowest frequency were likely to be the least successful foragers but they lived, on average, five days longer than spiders that decorated their webs at moderate or high frequencies. Spiders that decorated at a moderate rate lived almost twice as

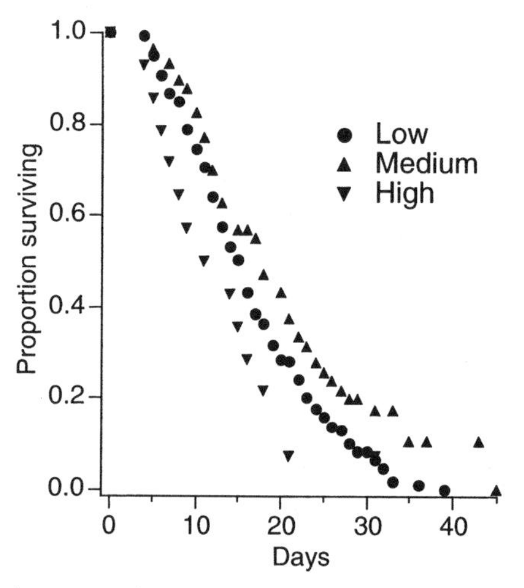

Figure 6.16 Survivorship among freely foraging spiders, grouped according to the frequency at which they decorate their webs, namely low-frequency decorators (<25% of all new webs were decorated; up-right triangles) medium-frequency (25–75% of all new webs were decorated; circles), and high-frequency decorators (>75–100% of all new webs; inverted triangles). The Kaplan–Meier analysis for comparing survivorship among individuals classified by treatment group showed that the mean lifetimes *m* for spiders that decorated their webs at high frequency (m = 12.6 days, SD =1.7, $n = 14$), medium frequency (m = 20.6 days, SD = 1.6. $n = 58$), and low frequency (m = 16.6 days, SD = 0.62, $n = 183$) differed significantly (Log-Rank, $\chi^2 = 9.0$, $P = 0.01$; Wilcoxon, $\chi^2 = 7.0$, $P = 0.03$). (From Craig et al. 2001; reprinted with permission from *Evolution*, Allen Press.)

long as spiders that decorated frequently, and four days longer than spiders that decorated only rarely. In fact, only seven out of 255 spiders (3%) survived more than ten days if they decorated their webs at high frequency (Craig et al. 2001). These results support the testable hypothesis that spiders that carry a high-decorating phenotype and forage in favorable environments will grow faster and reach reproduction sooner than spiders that do not.

The conclusions drawn from the results of the field and laboratory experiments described in this and in the two previous chapters can be summarized as follows:

1. Silk color renders webs either visible or invisible, depending on the contrast that it produces against the background.
2. Visibility or invisibility of spider webs is based on the insects' visual capacities.
3. The same holds true for spatial parameters of the web, particularly for patterns of web decoration.
4. The advantage of making webs visible is that they attract prey. The disadvantages are that they also attract the spiders' predators and that insects may learn to avoid the webs.
5. The frequency of web decorating affects the life span of spiders. The high cost of web-decorating behavior and, at the same time, the high reproductive payback of lucky decorators render the evolution of this behavior particularly sensitive to local ecological conditions.

These findings suggest that insects have had a selective effect on how spiders use silks and possibly also on the evolution of silk reflectance properties that are the result of silk structure. Spiders spin silks to serve specific functions correlated with the ability of prey to see webs and also with some preferences that are innate to insects. For example, webs are not decorated with just any type of silk, but in the case of *Argiope*, aciniform silk that shows a spectral peak in the UV region—a reflection to which insects are spontaneously attracted. Furthermore, web decoration patterns seem to mimic flower patterns and nectar guides. Although it is quite possible that some of the silk properties discussed here have evolved independently of insects' visual capabilities, it is clear that the properties of silk itself (see also chapter 3) have evolved due to the occurrence of insects, namely, for the sake of catching and retaining them.

Thus, for survival, spiders cannot dispense with silks. However, secreted silks and fibrous proteins are metabolically "expensive." The next chapter will deal with the cost of fibrous proteins and silks and its potential role as a selective factor in the evolution of the large diversity of silk proteins.

7

Inter-Gland Competition for Amino Acids and the ATP Costs of Silk Synthesis*

Silks are induced, highly expressed proteins that are secreted and used externally by insects and spiders. The numbers of ways in which silks are used and the ease of obtaining resources necessary to synthesize them affect the metabolic cost of silk and hence the evolution of the arthropods that produce them. For example, the silk systems of ancestral, nonaraneomorph spiders are limited and different silk proteins seem to be used nonspecifically (see chapter 3). Nonaraneomorph foraging systems are also limited, as is the metabolic efficiency of ancestral spiders (two pairs of book lungs, no tracheae). In contrast, the metabolic systems of Araneoidea are more efficient (one pair of book lungs, tracheae) and their silk systems and foraging behaviors appropriately complex. The single pair of silk glands of herbivorous moth larvae makes them seem relatively simple compared to those of the araneomorphs. Just prior to metamorphosis, however, the volume of silk that moths produce is enormous. Such striking differences in the diversity of silk systems, the volume of silk produced and the life histories of silk producers (that is, herbivores versus predators, hemimetabolous versus holometabolous development) suggest that the evolution and complexity of silk secretion systems may well be correlated with the abundance and predictability of available resources. In this chapter, I will explore the relative costs of

*Parts of this chapter are reprinted from Craig, et al. 1999 with permission from Elsevier Science B.V.

silks produced by insects and spiders and the possible implications of cost factors to silk systems' complexity and their molecular genetic organization.

The amino acids organisms synthesize are those needed in large quantities and on a predictable basis

The sum of the allocation of resources to different metabolic functions is often referred to as an organism's energy budget. An energy budget includes two measures: (1) the cost of metabolite synthesis; and (2) a measure of how much of the metabolite is used. Assuming that energy availability is not infinite, the amount of energy that can be diverted to an organism's specific needs—its development, growth, and reproduction—must be balanced. If the amino acids needed to make a protein are abundantly available, synthesis of the protein may not be metabolically costly. If the organism must invest a large amount of energy into synthesizing or obtaining specific amino acids from its environment, however, the metabolic cost of the protein will be high. This chapter is a comparative analysis of the production costs (invested ATP) of amino acids that make up the cost of silk produced by herbivores and predators. It does not address the cost of energy allocation to silk versus other organismal needs, i.e. the "distribution" costs.

Underlying these analyses is the assumption that the amino acids that arthropods synthesize must be those needed in the highest volumes and on a predictable basis. Organisms such as *Escherichia coli* are able to synthesize all the amino acids they need. However, when an amino acid becomes abundant in the animal's environment, it stops synthesis to harvest it (Neidhardt et al. 1990). This makes more energy available for other functions and to synthesize other amino acids that the animal needs. We propose that the synthesized amino acids are those the organism needs in most abundance, and that it cannot rely on the environment to provide. Contrary to current terminology, the synthesized amino acids are better thought of as the "essential" amino acids since they are needed in greatest abundance. The "nonessential" amino acids can be obtained with predictable frequency from the environment. This focus on the synthetic costs of amino acids can be related directly to the molecular genetic cost of the proteins. A second level of analysis, but one that is not explored here, would focus on the cost of the nucleic acids that code for silk. An analysis of the effects of cost selection could lend insight to the organization and evolution of genomes and their proteins.

The central metabolic pathways provide a common currency, ATP, through which the costs of protein synthesis can be compared

All organisms use the same metabolic pathways to break down ingested material and produce the energy (ATP) needed for protein synthesis (figure 7.1). These pathways include the Embden–Meyerhof pathway, which converts glucose 6-phosphate to pyruvate; the tricarboxylic acid cycle (TCA), which oxidizes acetyl CoA to CO_2; and the pentose phosphate pathway, which oxidizes glucose 6-phosphate to CO_2. The intermediate metabolites of these pathways provide starting molecules for the synthesis of all of the amino acids that are produced by insects and spiders. Therefore, the amount of ATP devoted to amino acid synthesis can be calculated as the net energy that is sacrificed when a metabolite has been diverted from the breakdown of glucose plus the energy invested in its particular synthetic pathway (Craig and Weber 1998).

With the exception of a unique pathway by which glycine is synthesized by Lepidoptera, the synthetic pathways by which amino acids are produced are thought to be similar for insects and spiders (Greenstone 1979). The amount of ATP invested in the production of amino acids represents a common currency through which a relative investment in proteins, and hence silks, can be analyzed. Therefore, comparison of the composition of silks pro-

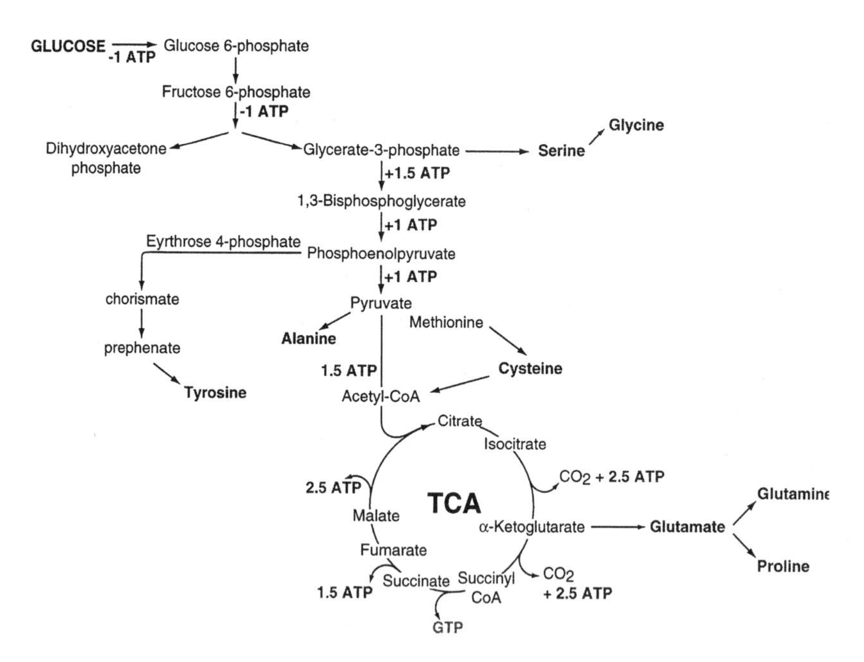

Figure 7.1 The baseline cost for amino acid synthesis can be estimated from the metabolic pathways.

duced by insects and spiders will reveal the inherent metabolic costs of amino acids diverted from the central metabolic pathways for the synthesis of silk.

Methods to estimate silk cost

We compiled available data (figure 7.2) on the amino acid composition of silks produced by Lepidoptera in the Psychidae, Galleriinae, Lasiocampoidae, Bombycoidea, and Noctuoidea; by Hymenoptera in the Argidae and Tenthredinidae (sawflies, Lucas and Rudall 1968) and in the Apidae, Vespidae, and Sphecidae (bees and wasps: Lucas and Rudall 1968; Espelie 1990; Espelie 1992; Singer 1992); by the Coleoptera *Hypera postica* (Kenchington 1983); by the Mantoidea *Mantis* sp. (Rudall and Kenchington 1971; Stehr 1987); by the Neuroptera *Chrysoptera flava* (Kenchington 1983); by the Diptera *Chironomous tentans* (Case 1994); by the Triochoptera *Pycnopsyche guttifer* (Case 1994); and by spiders in the genera *Araneus* and *Nephila* (Lucas and Rudall 1968). We supplemented these data with an analysis of silks collected from two different genera of larval Lepidoptera, one species of Embiidina, and seventeen species of spiders. Amino acid analyses of the silks were carried out on a Waters-Pico-tag system. All fibers were hydrolyzed in HCl, neutralized, and derivatized with phenylisothiocyanate. The derivatized amino acids were separated and quantified by reverse-phase HPLC.

The total cost of any amino acid can be equated to the energy (in the form of ATP) that is lost whenever a metabolite is diverted from the oxidation of glucose, plus the net energy invested into its specific synthetic pathway. Table 7.1 lists the estimated ATP costs of the input metabolites, the electron carriers, recycled by-products, and direct ATP investments that result when insects and spiders synthesize the amino acids found in the proteins they express.

Comparative approaches and available data

The amino acid distributions of silks and fibrous proteins are plotted on phylogenies for the orders Hymenoptera, Lepidoptera, and Araneae to determine which taxa might be useful for extended phylogenetic contrast (figure 7.3A–C). Some taxa are included in the phylogenies to clarify their ancestral-descendent relationships but for which silk data were not available. Cross-taxa comparisons of silks are limited to those derived from similar glands that are produced during comparable stages of development and that are used for similar purposes (see chapter 1). For example, we compared the composition and cost of silks produced in the colleterial glands by female mantids (*Mantis* sp.: Dictyoptera: Mantoidea; Rudall and Kenchington 1971; Stehr 1987) to the average cost of silks produced

Table 7.1 Estimated bioenergetic costs of amino acids produced by spiders, Lepidoptera, and other insects. (From Craig et al. 1999.)

Amino acids	Metabolites*	By-products (energy recycled)	ATP**	NADH 1.5 ATP	NADPH 2.5 ATP	Energy sacrifice (ATP)***
Alanine (A)	pyr, αkg	αkg			1	15.0
Aspartate (ASX)	oaa, glu	αkg				16.0
Cysteine (T)	M, S	αkb	3			−1.5
Glutamate (Glx)	αkg				1	23.5
Glycine***(*G*)	glyx, A	pyr				14.5
Proline (P)	glu		1		1	27.0
Serine (S)	gly-3p, glu	αkg		−1		14.5
Tyrosine (Y)	F				1	2.5

* Metabolite data and costs extracted from Gilmour 1965; Bendor 1985; Stryer 1995.
** αkg, α-ketogluterate (21 ATP); αkb, α-ketobuterate (19 ATP); glu, glutamate (23.5 ATP); glycerate-3-phosphate, gly-3p (14.5 ATP); glyx, glyoxylate (12 ATP); iso, isocitrate (12 ATP); oaa, oxaloacetate (13.5 ATP); pyr, pyruvate (12.5 ATP); F, phenylalanine (0 ATP); M, methionine (0 ATP); S, serine (14.5 ATP); suc, succinate (6 ATP); ***spiders and insects other than Lepidoptera. As in other insect and spider taxa, alanine, rather than glutamate, serves as the chief amino-group donor in glycine synthesis; however, in the silk glands of Lepidoptera, alanine is an intermediate metabolite that reslts from the transamination between glutamate and pyruvate (Gilmour 1995). Therefore, the total cost to synthesize glycine in the silk blands of Lepidoptera is estimated to equal 8.5 ATP and results from 1 GTP plus 2 NADH required to produce glyoxylate via conversion of isocitrate to succinate, and 1 NADPH (2.5 ATP) required for the transamination between glutamate and pyruvate.

in colleterial glands of adults in the order Coleoptera (*Hypera postica*: Curculionidae; Kenchington 1983) and Neuroptera (*Chrysopa flava*: Chrysopidae; Lucas and Rudall 1968). All three groups of insects use silks to protect their eggs and produce them in similar glands during similar stages of their lives (Lucas and Rudall 1968; Stehr 1987; Sehnal and Akai 1990).

With the exception of a few taxa (for example Psocoptera), the only insects that produce labial gland silks are those whose development is holometabolous (chapter 1). These include larval Hymenoptera, Diptera, Siphonaptera, Trichoptera, and Lepidoptera. The estimated costs and amino acid distributions of silks spun by the Hymenoptera are mapped on a phylogeny (figure 7.3A) drawn from recent, independently derived molecular phylogenies (Dowton and Austin 1994; Vilhelmsen 1997; Schmitz and Morritz 1998). Even though both ancestral and derived Hymenoptera larvae produce silks (Stehr 1987), silks are issued from larval labial glands among the Symphyta, while they are produced in Malpighian tubules of adult bees and wasps among the Apocrita (Stehr 1987; Maschwitz et al. 1991; with the exception of the colleterial gland silks of the Colletidae (Apoidea) and the labial gland silks of larval weaver ants). Because these glands may not be homologous, within-order comparisons were not made.

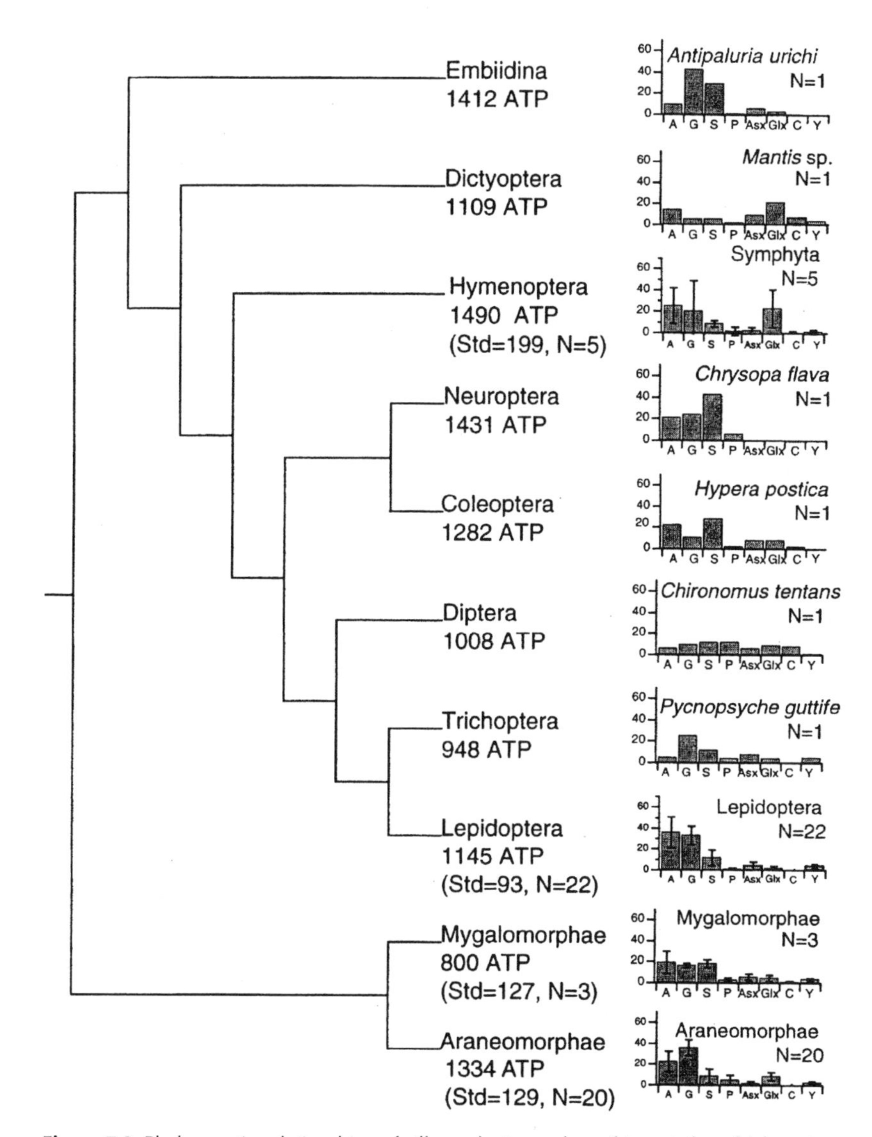

Figure 7.2 Phylogenetic relationships of silk-producing orders of insects for which amino acid data are available. Insect tree redrawn from Kristensen 1981. Nonparametric comparison of these distributions shows no statistical difference in distribution location. However, visual inspection of the data suggest that some silks are composed largely of alanine, glycine, and serine (Embiidina, Lepidoptera and dragline silks of spiders), while other silks are more varied (Mantoidea, Coleoptera, Triochoptera). (From Craig et al. 1999. Reprinted with permission from Elsevier Press.)

Substantial data are available only on silks produced by Lepidoptera and spiders and therefore these groups are the focus of our analysis. The Lepidoptera phylogeny illustrated in figure 7.3B is redrawn from a phylogenic estimate proposed by Minet that is based on morphology. It does not include all of the taxa that Minet analyzed (Minet 1994), and serves only to indicate the relative evolutionary relationships among the Ditrysia for which amino acid data are available. Furthermore, cross-taxa comparisons were made only among groups for which we could estimate mean costs and mean standard errors. For example, because we had only one sample of silk produced by larval Galleriinae, we compared the mean cost of silks produced by the Psychiidae with the mean costs of silks produced by Lepidoptera in the more distant taxa Lasiocampoideae. An additional comparison was made between the Bombycoidea (including Bombicidae and Saturnidae) and the Papilionidae.

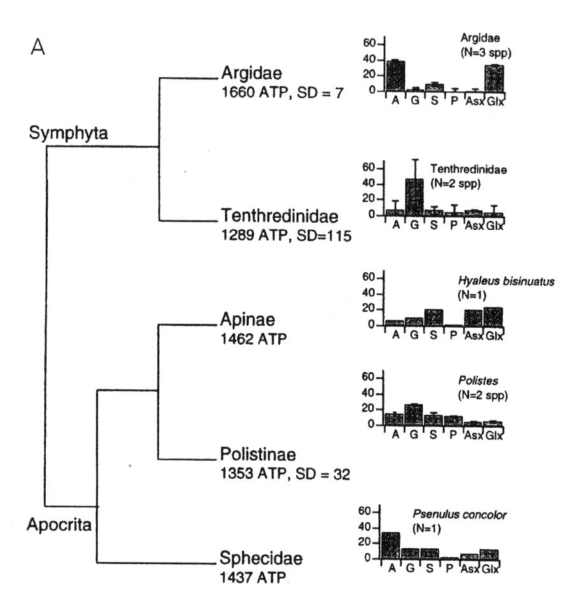

Figure 7.3 (A) Phylogenetic relationships among silk-producing Hymenoptera, (B) Lepidoptera, and (C) spiders for which amino acid data are available. Proposed phylogenetic relationships among the Hymenoptera are redrawn from recent, independently-derived molecular phylogenies (Dowton and Austin 1994; Vilhelmsen 1997; Schmitz and Morritz 1998). Branch support values not available. Hymenoptera species include *Pachylota audouinii, Digelansinus diversipes, Arge ustulata,* (Argidae); *Nematus ribesii, Phymatocera aterrima* (Tenthredinidae); *Psenuls concolor* (Sphecidae); *Hyaleus bisinuatus* (Apoideae). The location of the distribution of synthesized amino acids and estimated metabolic costs of silks produced by the ancestral (Symphyta) and derived (Apocita) Hymenoptera were not statistically different. (From Craig et al. 1999. Reprinted with permission from Elsevier Press.)

continued

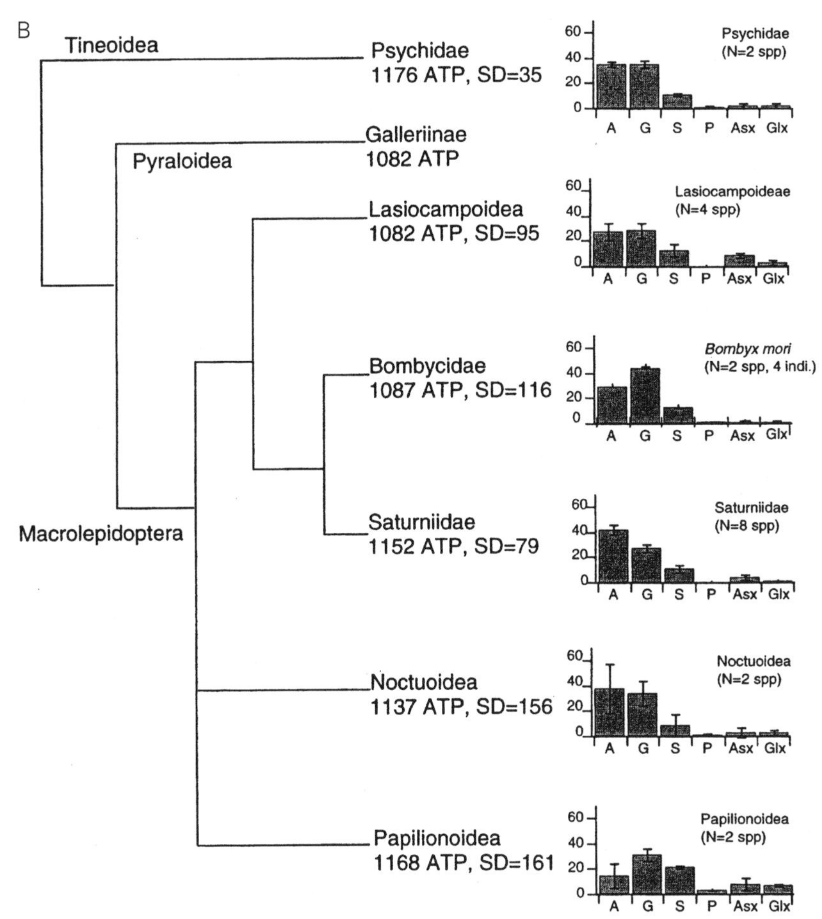

Figure 7.3 (*continued*) (B) The proposed phylogenetic relationships are redrawn from the independently derived phylogenies of Minet (1994) and are based on morphological characters and parsimona analysis. Branch support values are not available. The distribution of amino acids for silks produced by the Ditrysia show no statistical differences, nor do they differ statistically in estimated cost of silk synthesis. Lepidoptera species include: *Clania* sp., *Canephora asiatica* (Psychidae); *Galleria mellonella* (Galleriinae); *Lasiocampa quercus, Dictyoploca japonica, Pachypasa otus, Braura truncata, Malacosoma neustria* (Lasiocampoideae); *Antheraea assamensis, A. pernyi, A. mylitta, A. yamamai, Attacus pryeri, Cricula andrei , Eriogyna pyretorum, Loepa katinka* (Saturniidae); *Bombyx mori, B. mandalina* (Bombycidae); *Bena prasinana* (Agrotidae), *Arctia caja,* (Arctiidae), both Noctuoidea; *Jalmenus evagorus* (Lycaenidae), *Thisbe irnea* (Riodiniidae) both Papilionoidea. (From Craig et al. 1999. Reprinted with permission from Elsevier Press.)

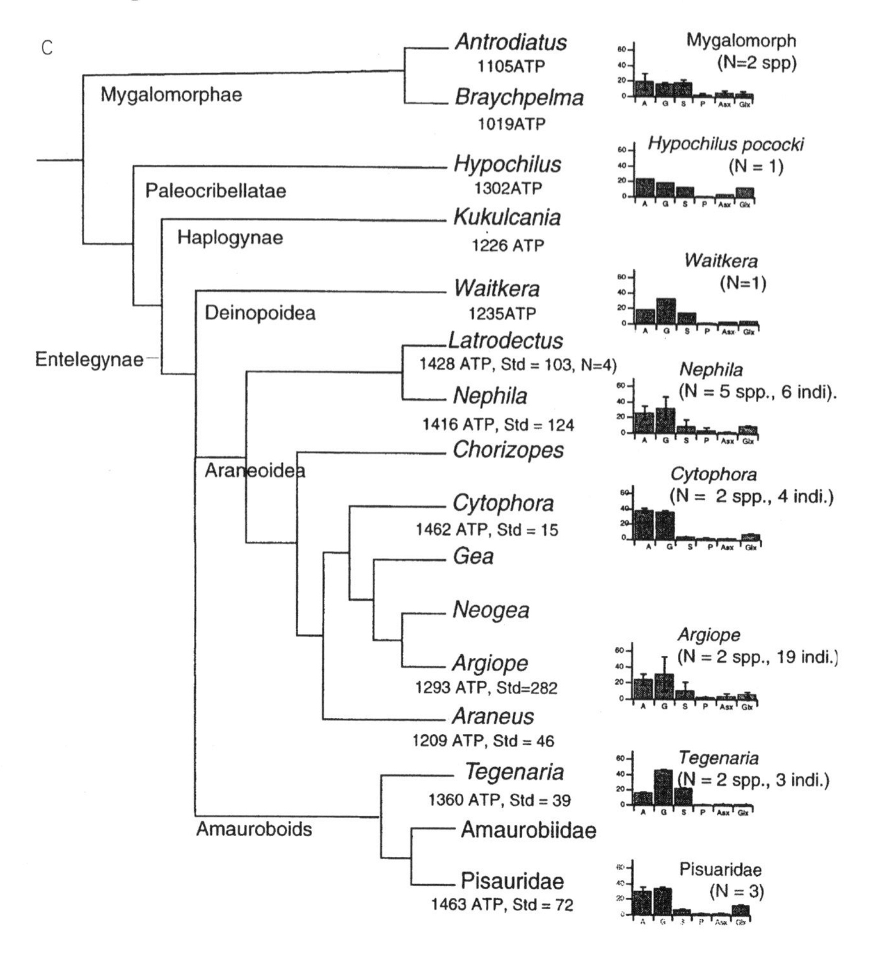

Figure 7.3 (*continued*) (C) Phylogenetic relationships are estimated from independently derived phylogenies of spiders (Scharff and Coddington 1997; Griswold et al. 1998) and are based on morphological data. Statistical analyses not available. Comparison among all taxa show that the locations of the distribution of amino acids do not differ significantly. However, the data suggest that the metabolic costs of silks produced by the Mygalomorphae are less costly to synthesize than the metabolic costs of silks produced by the Araneomorphae. The most striking difference among silks produced by the Araneoidea and other spider taxa is the high degree of variation in silk composition between species within some spider genera, as well as between individuals within some species. Spider species represented on the cladogram: *Antrodiatus apachecus, Brachypelma smithi* (Mygalomorphae); *Hypochilus pococki* (Hypochilidae), *Kukulcania hibernalis* (Filistatidae); *Waitkera waitkerensis* (Uloboridae); *Latrodectus nactaris maxiaus, Latrodectus* sp.(Theriididae); *Nephila clavipes, N. edulis, N. madagascariensis, N. tetragnathoides, N. plumipes* (Tetragnathidae); *Cytophora* sp., *Argiope argentata, A. appensa, Tegenaria* sp. 1, *Tegenaria* sp. 2, (Agelenidae), *Araneus diadematus, A. cucurbitanus, A. undatus.* (From Craig et al. 1999. Reprinted with permission from Elsevier Press.)

The evolution of silk glands and silk synthesis in spiders is even more complex than the evolution of silks in insects because individual spiders produce silks in multiple glands throughout their lives. The evolutionary relationships among taxa for which we had data on silk amino acid composition were estimated using recent phylogenies of the araneomorph spiders (Coddington and Levi 1991) and the Orbiculareae (Scharff and Coddington 1997; figure 7.3C). All of the silks sampled from the Orbiculareae are drawn from the ampullate gland (dragline silks) and are, therefore, homologous. The phylogenetic relationship among *Nephila* and spiders in the "argiopine" clade, however, remains controversial (as does the phylogenetic position of the Oribiculariae (C.E. Griswold, pers. commun.). The location of the distributions of amino acids spun by both Lepidoptera and spider taxa are compared, as are the estimated costs of the silks they spin. The silks from *Antrodiatus* and *Brachypelma* were collected from their nests and the specific glands in which they were produced are unknown (see chapter 3). These data suggest that the fibrous proteins produced by *Antrodiatus* and *Brachypelma* contain less alanine, glycine, and serine and more amino acids gathered externally than the *MA* silks produced by the more derived araneomorph species.

The amino acid compositions of silks spun by arthropods vary in proportions of alanine, glycine, and serine

The distributions of amino acids that make up the composition of silks produced by arthropods are summarized across phylogenetic groups (table 7.2) and illustrated in figures 7.2 and 7.3. Sorting silks on the basis of their amino acid composition reveals three groups: (1) silks with 60% or more of their composition consisting of two of the three amino acids alanine, glycine, and serine; (2) silks with 60% of their composition consisting of a combination of two of the amino acids alanine, glycine, and serine plus either proline or glutamine; and (3) silks with no two amino acids whose sum equals 60% or more of their composition. The small sample sizes available for most of these data prohibit statistical comparison of the distributions. However, the data suggest that almost all of the silks produced by Lepidoptera larvae fall into group 1. Silks produced by Embiidina, larval Symphyta, and spiders in the Oribiculariae characterize group 2. Silks produced by Diptera, Trichoptera, and Coleoptera larvae, adult mantids, adult Hymenoptera, and the mygalomorph spiders fall into group 3. Comparison of the locations of the amino acid distribution of silks spun by Lepidoptera and dragline silks spun by spiders, two groups for which adequate sample sizes exist, show that

Table 7.2 Comparisons of synthetic costs of silks between taxa by type of gland. While the limited data may suggest a trend towards reduced silk cost among organisms whose resource base is predictable (herbivores), trends in the cost silks produced by predatory spiders, whose resource base is likely to be variable, are more complex. (From Craig et al. 1999.)

Ancestral taxa	*Derived taxa*	Direction of difference between contrasted pairs
Contrasts among silks produced by herbivores (labial gland)		
Psychiidae 1176 ATP, Std=35, n=2 species, 2 individuals total	Lasiocampidae 1082, Std=95, n=4 species, 4 individuals total	+
Bombycoidea 1131, Std=116, n=5 species, 12 individuals total	Lycaenidae 1060, Std=20, n= 2 species, 2 individuals total	+
Symphyta (Hymenoptera) 1446, Std=184, n= 4	Lepidoptera 1117, Std=110, n=20	+
Contrasts among silks produced by predators and herbivores (colleterial gland)		
Apocrita 1232, Std=160, n=4	Neuroptera+Coleoptera 1251, Std=46, n=2 species, 2 individuals total	–
Contrasts among ampullate gland silks between predatory spiders		
Hypochilus (Paleocribellatae) 1302, ATP, n=1 species, 1 individual	*Filistata* (Haplogynae) 1226 ATP, n=1 species, 1 individual	+
Tegenaria 1360 ATP, Std=39, n=2 species, 3 individuals total	*Pisaura* 1463 ATP, Std=72, n=1 species, 3 individuals	–
Nephila 1416, Std=124, n=5 species, 6 individuals total	*Araneus* 1209 ATP, Std=46, n=2 species, 3 individuals total	+
Cytophora 1462, Std=15, n=2 species, 4 individuals total	*Argiope* 1293, Std=282, n=2 species, 18 individuals total	+
Waitkera 1235, n=1 species, 1 individual	*Latrodectus & Nephila* 1421, Std=110, n=7 species, 9 individuals total	–

they differ significantly (Mann–Whitney, $n = 16$, $P < 0.05$). These differences could reflect the cost of producing the silk or could indicate selection for different functions.

Baseline costs and compositions of silks vary between 1649 ATP and 96% synthesized amino acids (cocoon silks produced by *Pachylota audouini*, sawfly larvae, Hymenoptera) and 798 ATP and 59% synthesized amino acids (underwater nets produced by *Pycnopsyche guttifer*, caddisfly larvae, Trichoptera). If all the synthesized amino acids were equally costly to produce, the metabolic cost of a protein would vary directly with the proportion of synthesized amino acids it contains. While the data plotted in figure 7.4 show this relationship ($P \ll 0.001$, $r^2 = 0.32$), the scatter in the data reveals that variation in amino acid composition can affect total silk cost by as much as 22%. For example, 87% of the composition of dragline silk produced by the spider *Tegenaria* sp. 1 consists of alanine (16 ATP) and serine (12.5 ATP), yielding a total estimated cost of 1360 ATP (STD = 39). In contrast, 87% of the silk produced by the sawfly larvae *Pachylota audouinii* is made up of synthesized amino acids, but only 50% of these are alanine, serine, and glycine, yielding a total cost of 1662 ATP. Glutamine, costing 24 ATP per amino acid, makes up the remaining 36% of the protein; as a result, the metabolic cost of silks produced by *P. audouinii* is estimated to be 22% higher (over 300 ATP) than the metabolic cost of silks produced by *Tegenaria* sp. 1. Such substantial differences in costs imply that structural constraints or selective constraints demand that higher cost amino acids be used.

Direct comparison of amino acid costs suggests that dragline (*MA*) silks produced by araneomorph spiders are more costly than cocoon (*Fhc*) silks produced by herbivores

Using this approach, the costs of silks produced by herbivorous insects and predatory spiders can be calculated and compared. Figure 7.4 shows that the costs of spider dragline silks appear to be significantly higher than the cost of silks produced by larval Lepidoptera (glycine = 8.5 ATP for Lepidoptera, glycine = 14.5 ATP for spiders; ANOVA, $P < 0.0001$, $n = 42$). In fact, the silks produced by Lepidoptera larvae are less costly than a random assemblage of amino acids. This is perhaps not surprising because 97% of the silk produced by herbivorous, Lepidoptera larvae is composed of alanine, serine, and glycine (estimated 16, 14.5, and 8.5 ATP per amino acid, respectively) and the mean cost of synthesizing any amino acid is equal to 13.3 ATP (STD = 9.5). Alanine, serine, and glycine are also the major components of the silks produced by spiders, but substantial amounts of glutamine and proline (10% or more) differentiate spider silk costs from silks produced by the Lepidoptera (figure 7.3). As a result, dragline silks produced by the Araneoidea require significantly more ATP than silks produced by larval Lepidoptera.

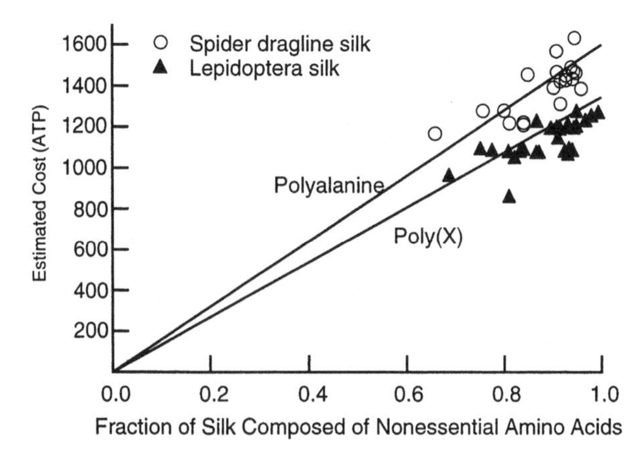

Figure 7.4 Estimated synthetic costs of silks are plotted as a function of their proportion of nonessential amino acids. While the silk data shows that cost increases with proportion of amino acids synthesized ($P < 0.001$, $r^2 = 0.32$, $n = 42$), the scatter in the data reveals that variation in amino acid composition can affect total silk cost by as much as 22%. Silks produced by herbivorous Lepidoptera larvae are slightly less costly than silks composed of a random assemblage of amino acids, while silks produced by spiders are slightly less costly than a fiber composed entirely of alanine, the putative ancestral polypeptide of spider silk fibroins. (From Craig et al. 1999. Reprinted with permission from Elsevier Press.)

Comparison of *MA* silk produced by ancestral and derived species suggests a trend toward reduced silk costs among the cribellates and between the *MA* and *Flag* silks of the Araneoidea

Data for the estimated costs of silks produced in homologous glands of ancestral and derived arthropods are limited and hence cannot be rigorously compared. Nevertheless, viewing cost data in a phylogenetic context points to taxa where more detailed analyses could be fruitful. For example, the mygalomorph spiders *Antrodiatus apachecus* (Antrodiaetidae) and *Brachypelma smithi* (Theraphosidae) produce fibrous proteins characterized by a mean estimated cost of 1062 ATP (STD = 60). The *MA* silk (first evolved silk fibroin) produced by the paleocribellate spider *Hypochilus pococki* (Hypochilidae; most ancestral of the araneomorph spiders), requires a higher ATP investment, approximately 1302 ATP, than the fibrous proteins produced by the nonaraneomorphs. *Waitkera waitkerensis*, a derived cribellate spider that spins an orb web, produces dragline silk that costs approximately 1235 ATP. *W. waitkerensis* is a member of the superfamily Deinopoidea, the sister outgroup to the Araneoidea. These silks are less costly than the *MA* silks produced by the araneoids.

Comparison of *MA* silks produced by Araneoidea is less clear than the preliminary results for the cribellate species. In some cases, costly

MA silks are produced by more derived taxa and in some cases they are not. The theridiids (*Latrodectus*, 1428 ATP, STD = 103) and tetragnathids (*Nephila*, 1421, STD = 110), which represent ancestral araneoids, produce *MA* silks that are more costly than the *MA* silks produced by the cribellate spiders. Comparing the hunting spiders *Tegenaria* sp. 1 and *T.* sp. 2 (both Agelenidae) with *Pisaura* shows that *MA* silk produced by the ancestral species is less costly (1360 ATP, STD = 39) than the silks produced by the more derived *Pisaura* (1463 APT, STD = 39). As the data currently stand (and they are admittedly weak), they suggest a trend to reduced costs in *MA* silk among web spinners but not hunting spiders. Deeper-level comparison between the cribellate *MA* and the Orbiculareae *MA*, however, shows an initial cost increase in *MA* silks followed by cost decrease.

The proposed evolution of spider silk costs is more complex than the direction of cost changes observed when ancestral-descendent comparisons are drawn between the Lepidoptera (table 7.2). The most recently evolved silk fibroin, *Flag* silk, is less costly than *MA* silks. Cylindrical gland silk, however, is more costly than *MA* silk and used for the same purpose as lepidopteran cocoon silk. The fact that ancestral and derived silk glands do not follow a clear evolution to reduced silk costs suggests that other selective factors are important.

Spider silk glands may have evolved through intra-gland competition for amino acids that the spiders synthesize

Ancestral spiders (Mesothelae) produce fibrous proteins in acinous-like glands that are morphologically similar but that have different intra-gland environments. The spiders use the proteins indiscriminately, however, regardless of composition (Kovoor 1987). Such indiscriminate use of the proteins suggests that the primary selective force on early silk gland systems may have been simply to produce a volume of fibrous material that could be secreted (table 7.3). In this scenario, new glands and new processing systems could be the result of inter-gland competition for limited amino acids. One might imagine that when ancestral species began to use silks to retain prey (and subsequently forage above ground), spiders were able to increase their rate of prey intake. In the new, above-ground selective environment, the thread-like fibrous proteins were spun into funnel retreats and sheets (chapter 3). Not only is the volume of material needed for sheets and above-ground funnels greater, but also one could expect that proteins themselves would be under selection for new properties. The new selective regimes would favor a shift from fibers that only bind soil to fibers that provide protection in the absence of other shelter.

Table 7.3 Proportion of synthesized amino acids in silks produced in different glands of *Araneus diadematus*. The silks proposed to have appeared early in spider evolution (piriform) and the fibrous proteins (aggregate, aciniform), contain a larger proportion of amino acids gathered externally than the fibroin silks.

Protein type (listed in possible order of evolution	Gland	Proportion of synthesized amino acids
Fibrous protein glue	Piriform	59.1
Fibrous protein glue	Aciniform	63.4
Fibrous protein glue	Aggregate	60.5
Fibroin thread	Major ampullate	76.7
Fibroin thread	Minor ampullate	84.3
Fibroin thread	Cylindrical	94.5
Fibroin thread	Flagelliform	92.8

The larger volume of silk that is produced by ancestral, above-ground foragers makes greater use of amino acids that spiders can synthesize. The significance of a shift from silks made up of gathered amino acids to synthesized amino acids is that it allowed derived spiders to be, in a sense, freed from the specific environments where predictable amounts of those amino acids might be found. The shift from use of gathered to synthesized amino acids in fibrous proteins would also have affected silk organization. Recall that the amino acids most accessible from the breakdown of glucose (but not necessarily the most expensive) are glycine, alanine, and serine, the three amino acids with the most compact side chains. These same amino acids are those that align into cross-linked, beta-pleated sheets when the silk is sheared or spun (see chapter 1).

The evolution of *MA* silks, which are spun silks, correlates with the evolution of 97% of all spider species (see chapter 3). Also correlated with spider diversification is the evolution of more efficient respiratory systems. The Mesothelae, the Orthognatha, and the ancestral family Hypochiliidae all have two pairs of book lungs. Oxygen carried by the hemolymph diffuses in the spiders' muscle across only about 60–90 μm (Foelix 1996). Most of the araneomorphs, however, have one pair of lungs; the second pair is modified into tubular trachae. While different families of spiders have differently developed tracheae (Opell 1987), those with highly branched tracheal systems that extend throughout the spider's body have more efficient gas exchange and hence respiratory metabolism. It seems unlikely that the complex silk systems of derived species could have evolved without a correlated evolution of spider respiratory physiology.

Gene organization that allows selective expression and/or selective editing of proteins may allow spiders to reduce silk costs during periods of food stress

Molecular sequence and gene regulation are studied intensively to elucidate the evolution of proteins. Less frequently considered, and often more difficult to determine, is whether factors such as available resources or physical environments can bias the regulation of gene expression. In chapter 2, I argued that if silks are produced from more than one protein (*MA* silks produced by *Nephila*) or if the structure of the mRNA transcript allows easy editing, spiders may more easily adapt proteins to a new physical environment. These same arguments can be made in light of silk costs. Figure 7.5 shows estimated costs for the primary repeating units of *Ma1*, *Ma2*, and *Bm-fhc*. Recall that the two components of *MA* silks have different repeating units whose ratios may vary. Figure 7.5 compares cost data for the two units, showing that GPGQQ, the primary component of *Ma1* and thought to result in fiber stiffness, is more costly to produce (average of 21 ATP per amino acid site) than GPGXX (average of 14 ATP per amino acid sites), the primary component of *Ma2* and the module attributed to fiber flexibility. In addition to its lower baseline costs, the GPGXX module has two variable sites. One of the amino acids is threonine, –1.5 ATP, which actually adds energy to the synthetic system. Another amino acid, valine, comes from the environment. The most costly amino acid contributing to the *Ma2* is glutamine, estimated to require 23.5 ATP. In contrast, the cost of GPGQQ is fixed and high due to the two units of glutamine (Q) it contains. In the case of food stress, a relative increase in the proportion of the *Ma2* protein and hence drop in the proportion of the *Ma1* component

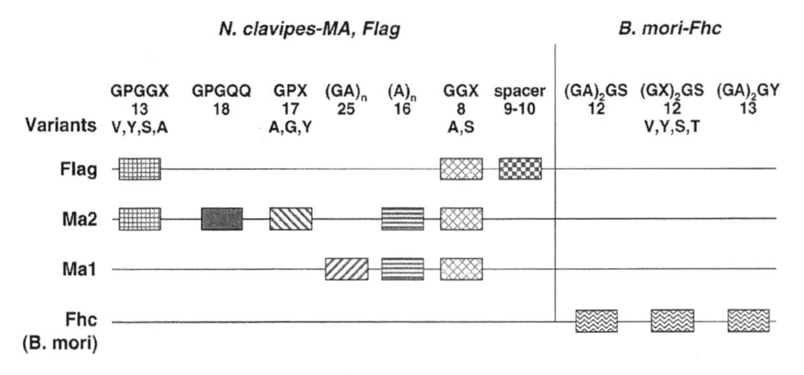

Figure 7.5 The metabolic costs of the major repeating units of *Flag* silks are less than any other fibroin silk produced by spiders. The costs of *Fhc* silk produced by *B. mori* are lower than any of the silks produced by spiders. (From Craig et al. 1999.)

would result in a lower cost for dragline silks, making them more similar in cost to *Flag* proteins. Such an event would also result in greater elasticity.

The ability to recycle silks allows the araneoids to reduce the metabolic costs of producing silk

Even though our understanding of spiders and silk systems is limited, the preliminary data presented in this chapter give every indication that, as in all other organisms, cost is a primary factor in spider evolution. The cost of web silks to cribellate spiders and nonweb-spinning spiders is particularly high because once the protein is spun, the spiders are unable to consume it and break down the silk threads for reuse. The araneoids, however, spin webs daily and their high rate of protein output is achieved due to their ability to consume and recycle their silk (Opell 1998). As much as 90% of the protein invested can be recouped (Peakall 1971).

The evolution of above-ground foraging and complex silk systems seems to have allowed spiders to disperse and track insects in virtually all of the environments in which they are found— terrestrial, aerial, and underwater. Despite all of their advantages, however, the production and secretion of silks is costly and the structural and functional constraints on silk proteins may limit potential cost reductions. One effect of spiders' maintenance of a high-cost silk system may be that it precludes the diversity of lifestyles and developmental patterns found in so many of the insects.

The most energy-efficient and ecologically numerous arthropods are those that live in caste-based societies that have achieved higher eusociality. There are many different "grades" of social systems that result in common nests, cooperative foraging, and cooperative reproduction. The pinnacle of social evolution, however, is the eusocial society, where a single queen takes over the reproductive tasks for an entire colony. Members of the colony, usually the same genotype as the queen, can then specialize in caring for the offspring, foraging, or colony defense and still be assured of genetic representation in future generations.

Why have spiders never evolved higher eusociality? The primary selective force driving eusociality, at least according to some authorities, is resource based (Wilson 1971). The eusocial and cooperative lifestyles of insect societies are much more energy efficient than is possible for the type of lifestyle achieved by spiders. If the primary selective force on eusocial evolution is energy and resource based, the high energy stress resulting from silk production would seem to select for a eusocial lifestyle. In the next chapter, I will explore possible constraints that may have limited the evolution of eusociality in spiders.

8

A One-Dimensional Developmental System and Life-Long Silk Synthesis May Preclude the Evolution of Higher Eusociality in Spiders

Sociality—cooperative nest building and the sharing of tasks—has evolved in all higher orders of insects (Hölldobler and Wilson 1990). Of these higher orders, the Hymenoptera (ants, bees, and wasps) and Isoptera (termites) are remarkable for the variety of social systems they include. Their life histories vary from solitary, to "primitively" social (individuals living in colonies with primitive caste systems; that is, they differ in physiology and size), to highly "eusocial" (individuals living in colonies characterized by reproductive and worker castes). The highest levels of social complexity are characterized by the evolution of queen and worker that includes the evolution of multiple, morphologically diverse worker castes (Wilson 1971; Wheeler 1986).

The most important advantage of sociality is "energetic": individuals living as a group are better able to withstand the effects of fluctuating environments. Whether resources are abundant or scarce, foraging is more efficient with group cooperation, and groups are better able to defend the nest and its offspring. Attesting to the advantage conferred by sociality is the fact that bees and wasps make up more than 75% of the total insect biomass and ants and termites alone make up about one-third of the entire biomass of the terrestrial, Amazonian rain forest (Hölldobler and Wilson 1990).

In spiders, however, social behavior has evolved in only about three dozen out of an estimated 100,000 spider species (Buskirk 1981; D'Andrea 1987). Furthermore, with the exception of the Australian huntsman spiders (Sparassidae), (Rowell and Avilés 1995), communal living has evolved only among spiders that spin webs (Krafft 1979; Buskirk 1981; D'Andrea 1987; Avilés 1997). The reason that cooperative living has evolved primarily in web spinners is thought to reflect the fact that: (1) a web provides a means of communication among the spiders (Shear 1970); and (2) silk production alone provides an incentive for sharing a nest (Riechert 1985). While there are some disadvantages to communal living (that is, some social spider groups attract parasites; Griswold and Meikle-Griswold 1987), the low frequency of sociality in spiders is surprising because spiders that live communally produce less silk per individual (Riechert 1985; Tietjen 1986; Jakob 1991), survive longer (Vollrath 1982; Riechert 1985; Avilés 1992), and catch larger or more prey (Nentwig 1985; Pasquet and Krafft 1989; Rypstra and Tirey 1989; Rypstra 1990). Moreover, communal living would seem to be at least a partial solution to the challenge presented by the energetic cost of silk production in spiders and the apparent conflict between energy allocation for silk and energy allocation for reproduction (see chapter 7). Nevertheless, the most advanced spider colonies are simply perpetually inbreeding colony lineages that give rise to daughter colonies or that become extinct without mixing with one another (Avilés 1997). Thus, while some web-spinners benefit from social living, the benefits are limited.

The lack of eusociality in spiders, despite their apparent "attempts" at sociality, led me to explore the selective factors that influence the evolution of complex societies in insects. In particular, I wanted to identify the primary selective factors affecting eusociality in ants and termites. I chose these two insects because ants are like spiders in that they are predators but unlike spiders in that they have holometabolous development (complete metamorphosis); termites are like spiders in that they are diploid and have hemimetabolous development (incomplete metamorphosis) but unlike spiders, they are herbivores.

Multiple selective factors favor the evolution of eusociality

Multiple selective factors favor the evolution of eusociality, including a defensible nest or food site, the potential for extended periods of development, monogamous reproductive pairs, high chromosome number (hence reduced variance in shared genes), inheritance of resources, and cooperative defense against predators, parasites, and competitors (Thorne 1997). One potentially significant factor influencing the evolution of eusociality in the Hymenoptera is their haplo-diploid system of sex determination. In Hymenoptera, all females are diploid (emerge from fertilized eggs) and all males are haploid (emerge from unfertilized eggs). Therefore, female offspring share, on average, one-half of their mother's genes. Because their father is haploid, however, female offspring share all of their father's genes. Hence, sisters have more genes in common with each other (the coefficient of relationship is 3:4) than they do with their mothers (the coefficient of relationship is 1:2). As a result, daughters that stay in the nest and care for their younger sisters, thereby contributing to their survival, are likely to have a greater impact on the species gene pool (due to the high degree of relatedness among sisters) than would offspring that risk founding their own colony and producing daughters to which they have a lower degree of relatedness (Hamilton 1964).

Although high genetic relatedness among offspring gives a special advantage to the evolution of group living, eusociality has also evolved in diploid animals. While a haplo-diploid genetic system of sex-determination lends an animal a clear propensity to evolve eusocial systems, it is neither necessary nor sufficient by itself. In diploid animals, including termites (reviewed in Thorne 1997), aphids (Aoki 1977; Aoki 1982), beetles (Kent and Simpson 1992), naked mole rats (Alexander et al. 1991), and tropical reef shrimp (Duffey 1996), the energetic advantage gained by sharing resources and defense—an advantage accruing to all colony members—may be the primary selective factor favoring eusociality.

The importance of a flexible developmental pathway

All eusocial species are marked by reproductive and worker castes that are behaviorally and physiologically distinct from each other (Wheeler 1991). A flexible developmental pathway and the physiological ability to reprogram this pathway in response to the environment are the most important preconditions for the evolution of eusociality (West-Eberhard 1989). For example, all termites in the ancestral genus *Kalotermes*—kings, queens, soldiers, workers, and alates (winged offspring)—retain the ability to reproduce as adults as well as considerable developmental flexibility prior to maturation. In addition, the developmental pathway taken by *Kalotermes* larvae is indeterminate at hatching. However, after the first molt, individuals are set on either a worker or reproductive path. At the second molt, in an example of developmental flexibility in the extreme, the sexual nymphs—that is, those individuals now set to follow the reproductive developmental pathway—can regress to their previous stage of development; in other words, they can "back" molt and reset their developmental pathway to become workers (Noirot 1985; reviewed in Thorne 1997). In the higher termites, for example, the Termitidae, developmental pathways are even more complex, the castes are morphologically even more divergent, and the worker line is usually sterile (Watson and Sewell 1981). Therefore, the divergence of termite reproductive and worker castes and of those of other eusocial animals falls outside the realm of genetic differences in character traits. Instead, this divergence is determined by how development is controlled.

In ants, phenotypic variation among castes ranges from continuous size variation (Wilson 1953; Wheeler and Nijhout 1984) to complete dimorphism (that is, discontinuous size variation, or polyphenism) (Hardie and Lees 1985; Wheeler 1986; Sterns 1989; Wheeler 1991; Emlen and Nijhout 2000; Tschinkel 1988). The fire ants, *Solenopsis invicta*, are characterized by a primitive form of polyphenism. They may represent what could have been a first step in the evolution of more complex worker castes (Wheeler 1990). The mechanism of caste determination among the fire ants is unknown, but it occurs at two different stages in larval development. Larvae exposed to the developmental hormone called juvenile hormone (JH) during early stages in development grow larger than larvae not exposed to JH. Their large pupal size results in the emergence of a "major," or large, worker. A major worker will also result if larvae are exposed to JH at the fourth stadia, this exposure extends the larval growth period (allowing extended feeding), which also results in larger pupal sizes. Thus, the presence of JH at two specific times (one early in larval development and one late) during fire ant development is correlated with two possible critical sizes, or the sizes at which developing larvae take the first physiological steps initiating metamorphosis (Wheeler 1990). It is the environment's influence on the endocrine system that triggers the morphological changes in initially similar individuals needed to create the various castes that make up a eusocial community. However, if a developmental pathway susceptible to these triggers is absent, these changes cannot occur.

Such flexibility to extend the length of the developmental period is indeed absent in spiders. All of the other selective factors affecting the evolution of eusociality in diploid organisms also affect spiders, so it seems probable that the absence of the potential for extended development may account for spiders' failure to achieve eusociality. The number of stadia (growth periods between molts) it takes araneoids to reach sexual maturity is an evolutionary correlate of their adult size (that is, large spiders molt more times than small spiders). The length of time spent in each growth period is determined largely by food availability. High rates of food input accelerate growth rates within an instar and thus shorten the time between molts, enabling the spider to reach a reproductive physiology more rapidly. It seems that the developmental periods of well-fed spider nymphs can be shortened but are never extended. Furthermore, there is no evidence that spiders have the type of developmental flexibility that results in the resetting of critical size. Resetting critical size would enable some spiders to produce more generations per growth season and have a greater genetic input to the population than those growing more slowly. It would be interesting to compare the evolutionary effects of rapid growth with the evolutionary effects of eusocial societies on adaptation to changing environments.

Hormonal mediation of development in eusocial insects

The shift in larval growth period and time of sexual maturation of larvae, hallmarks of eusocial ants and termites, are mediated by a two-dimensional regulatory system consisting of ecdysone, which causes morphogenic change, and JH, which modifies the effects of ecdysone to allow extended growth. Therefore, despite colony members' genetically equivalent genotypes, different developmental pathways triggered by environmental factors result in extreme morphological diversity. Because different tissues are more or less susceptible to the influence of ecdysone and JH, the growth rates of the body parts of the developing larvae can be "uncoupled" from the baseline growth rate of the larvae (figure 8.1). This means that body parts are able to grow at different rates. For example, soldiers grow disproportionately large head and mandibles in comparison to castes where head growth has not been uncoupled from the ants, baseline, developmental pattern (Emlen and Nijhout 2000). The spatial and temporal sensitivity of developing tissues to ecdysone and JH places the evolution of eusociality squarely in the domain of endocrinology in the insects (Wheeler and Nijhout 1984; Wheeler 1986); but see Volny and Gordon 2002. But is there any other pathway to eusociality in arthropods?

The presence of JH has been convincingly demonstrated in all the insects, Crustacea, and Acari (ticks and mites). However, JH has not been conclusively identified in spiders. Nor has it been identified in spiders' ancestral taxa, the Xiphosura (horseshoe crabs) and Pycnogonida (sea spiders) (reviewed in Kaufman 1997; see figure 1.5), suggesting that JH evolved independently in the Acari and once again with the insects and Pancrustecea. Ecdysone initiates silk gene transcription in insects, and perhaps in the cylindrical glands of spiders. In insects, transcription is inhibited by the presence of JH. In spiders, at least in the *MA* gland, gene transcription is neurally initiated (Candelas and Citron 1981), suggesting that blood borne ecdysone (as well as JH) may not control the onset of silk gene

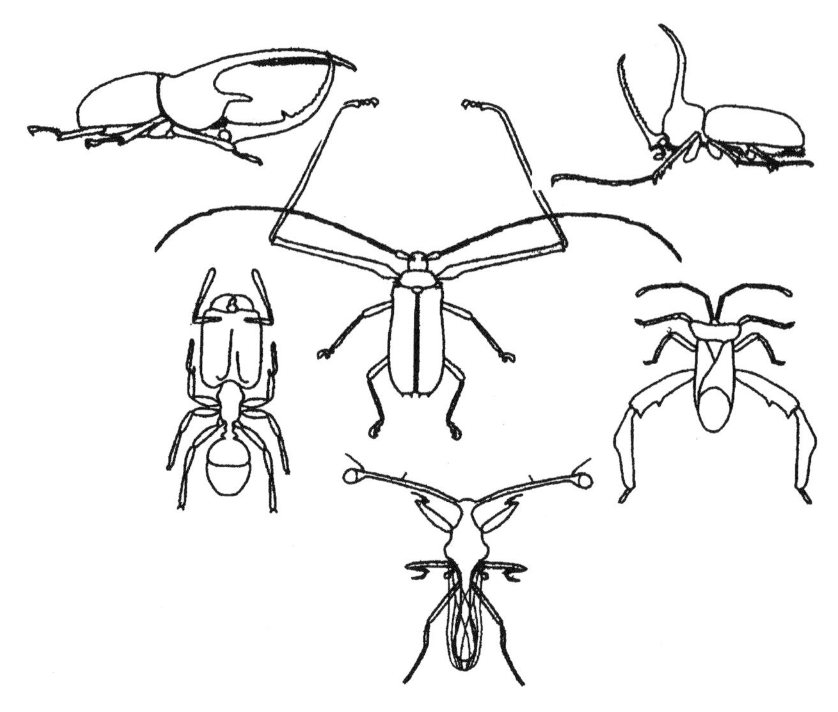

Figure 8.1 Exaggerated morphologies that can result from the uncoupling of growth patterns. Left to right, top row: head and thoracic horns in *Dynastes neptunus* (Coleoptera: Scarabaeidae); head and thoracic horns *in Golofa porteri* (Coleoptera: Scarabaeidae); *Middle row*: head width in *Phiedole tepicana* (Hymenoptera: Formicidae); forelegs in *Acrocinus longimanus* (Coleoptera: Cerambycidae); hind legs in *Acanthocephala declivis* (Hemiptera: Coreidae); *Bottom*: eyestalks in *Cyrtodiopsis whitei* (Diptera: Diopsidae). (After Emlen and Nijhout 2000; with permission from the *Annual Review of Entomology*, Volume 45. © 2000 by Annual Reviews www.annualreviews.org.)

transcription. Therefore, silk production in spiders may be largely independent of the hormonal controls of morphological growth. Does life-long silk production in spiders exclude the evolution of a generalized regulatory molecule such as JH?

Given this background of fundamental differences between the developmental pathways of eusocial insects and spiders, I pose the following questions:

1. Does the evolution of highly eusocial systems require the developmental hormone JH or is the JH control system simply the only control that has been identified? Are there other mechanisms of developmental control that could result in eusociality that have not been recognized or discovered?

2. Can studies of the silk-producing physiology of Lepidoptera provide any insight into silk production in spiders? If so, do the inhibitory effects of JH on silk synthesis, as demonstrated by larval Lepidoptera, preclude the evolution of eusociality in spiders, in which silk production is life-long?
3. Can reproductive and worker castes evolve in the presence of one regulatory hormone?

The fundamental assumption of this chapter is that by understanding the hormonal control of higher eusociality in insects, as well as the hormonal control of silk production in Lepidoptera, we may gain some insight to why higher eusociality has *not* evolved in spiders. To address this question, I have compared the primary developmental patterns and physiology of insects and spiders. In particular, the antagonistic roles of JH and ecdysone in morphogenesis and protein synthesis must be known. I hope to determine whether a lifelong commitment to silk production has, in some way, limited developmental flexibility in spiders. Furthermore, even if JH is absent in spiders, why cannot sufficient developmental flexibility be achieved through the evolution of multiple, hormone receptor proteins that are spatially and temporally variable? Is there a special advantage to having a hormone that modifies the effects of a primary regulatory protein?

Insects have three developmental pathways, but spiders have only one

For arthropods, the term "development" refers to the period from fertilization of the egg to the emergence of the adult. The majority of the Crustacea, the myriapods, and the ancestral hexapods are anamorphic (which is considered the ancestral developmental state for segmented animals), meaning they hatch without a full complement of body segments and appendages. Appendages are added later, during post-embryonic development and under the direction of ecdysone. Insect development is further divided into three general classes:

- direct development, in which hatchlings are fully formed but lack reproductive genitalia (wingless hexapods, for example, bristletails and silverfish);
- hemimetabolous insects (for example, cockroaches, grasshoppers, dragonflies, and true bugs), in which hatchlings emerge with a full complement of legs but continue morphogenic development while growing to sexual maturity; and

- holometabolous development, in which hatchlings emerge at an even less-developed stage and continue embryonic growth largely in the absence of morphogenic change until reaching pupation.

The holometabolous insects form a monophyletic group, but the hemimetabolous insects are polyphyletic (see chapter 1).

There are two hypotheses regarding the evolution of insect developmental patterns. One hypothesis is resource-based and suggests that the form the insect larva takes at hatching is a function of the amount of yolk provided by the egg; here, a diversity of hatching times results in a diversity of larval forms (Berlese 1913). Hemimetabolous insects have a large volume of yolk per egg; they therefore emerge after a relatively longer time in the egg and at a more advanced morphological stage—as nymphs—than do holometabolous insects (Berlese 1913; Truman and Riddiford 1999). With less yolk, holometabolous insects hatch as little more than freely foraging larvae, and the nymph stage is reduced to a single stage, the pupa. The second hypothesis proposes that there is no difference in the volume of yolk in insect eggs and that the nymph and larval stages are essentially the same. In this hypothesis, the pupal stage arose *de novo*. After examining the endocrine control of embryonic and post-embryonic larval development, it was proposed (Truman and Riddiford 1999) that the roots of metamorphosis are more likely derived from varying degrees of food resources stored in the egg (Hypothesis 1). Therefore, the evolution of the amount of time spent in the egg reflects the resources available to the embryos prior to hatching.

At first glance, spiders have hemimetabolous development: they emerge from the egg sac with a full complement of legs but are physiologically immature. Regrettably, there is limited information on the comparative life histories of spiders. We know little about spider post-embryonic development, developmental flexibility, and the hormonal control of development (but see Babu 1973; Babu 1975; Bonaric 1987; Bonaric and De Reggi 1987). We do know, however, that eggs of the same size, regardless of the phylogenetic status of the spider that produced them, contain the same amount of energy (Anderson 1990). Furthermore, spider egg size varies by two orders of magnitude and is directly correlated with spider size (large spiders produce large eggs; small spiders produce small eggs). This fact seems to suggest that the only developmental pathways open to spiders are via variation in size or some other unidentified mechanism. Furthermore, given the macroevolutionary trend to small size across the araneoids, spider development patterns seem to be constrained to a simple change in the number of developmental stages. The extended period of growth and uncoupling of growth patterns that enable eusociality do not seem to have evolved in spiders. Spiders—like hemimetabolous insects—do their primary morphological development in the egg; that is, they hatch with a full complement of body parts. Hence, at least on an ecological time scale, morphological development may be relatively unaffected by the environment. In contrast, development in eusocial insects is almost always determined solely on an ecological time scale, and the environment affects hormone production, which, in turn, affects changes in development.

Ecdysteroids regulate metamorphosis in the absence of JH

Viewing the evolution of eusociality through the lens of endocrinology makes it necessary to review briefly what is known about the major physiological factors that direct insect and spider development. Ecdysteroids (the term encompassing the various forms of ecdysone) regulate multiple processes during embryogenesis and development in the protostomians (Arthropoda, Echinodermata, Mollusca, Gastropoda, Annelida, Platyhelminths, Nemathelminthes, and Coelenterata) (Suzuki et al. 1984; Mercer 1985; Prudhomme et al. 1985; Bonaric 1987; Bückmann 1989; Horn 1989; Kremen and Nijhout 1989). They are sterol derivatives and considered to be phylogenetically "ancient" molecules (Franke and Käuser 1989, Käuser 1989). In evolution, ecdysteroids were first produced in the gonads and thought to have been used as pheromones to attract mates, and secondarily to have taken over control of some aspects of the cell cycle. Later in evolution, the brain became the major source of ecdysteroid production. By 1989, sixty-one different forms of ecdysone and their physiological effects had been identified (Rees 1989).

Neither insects nor spiders store ecdysone, the synthesis of which is regulated by a negative feedback loop (Watson et al. 1989; Nijhout 1994). In insects, ecdysone is secreted by the prothoracic gland when this gland is stimulated by the brain hormone prothoracotropic hormone (PTTH). When the hemolymph achieves a specific ecdysteroid titer, synthesis is repressed by the presence of ecdysone (Bidmon 1988; Richter et al. 1989). Spider endocrine glands and neurosecretory organs include Schneider's organs (neuroendocrine) and the Tropfencomplex (a neurohemal organ including the stomadeal bridge and esophageal ganglion) (Babu 1973; Gabe 1955; Gabe 1966). They are located in the abdomen in the Mygalomorphae and the prosoma in the Araneomorphae; they extend through both the abdomen and the prosoma in the ancestral Liphistiomorphae. Thought to be analogous to insect prothoracic glands and derived from the same cell lineage (ectodermal origin), the I Schneider's gland-Tropfenkomplex is the main site of ecdysone synthesis and release (Babu 1973). Histological studies show that the secretory activity of the I Schneider's gland-Tropfencomplex correlates with ecdysis (Babu 1973) and hence the titer of circulating ecdysone.

Ecdysone coordinates the physiological process of molting and metamorphosis for all arthropods. Production is confined to two discrete periods that are differentiated by the concentration of ecdysone in the hemolymph. Ecdysone concentration peaks prior to ecdysis and declines during molting, and production is resumed at a low background level after ecdysis (Sehnal and Akai 1990). The morphogenic role of ecdysone is controlled by the timing of expression and sensitivity of ecdysone receptor proteins. Specifically, different receptor isoforms are expressed in different tissues and are sensitive to ecdysone for only brief periods of time (Fujiwara et al. 1995; Jindra et al. 1996; Emlen and Nijhout 2000). In addition, the pattern of ecdysteroid expression fluctuates to control events at different downstream response pathways (Emlen and Nijhout 2000).

Low background concentrations of ecdysone are necessary for normal tissue development. However, prior to ecdysis, ecdysteroid concentrations peak to initiate arthropod molt (Bonaric and De Reggi 1987; Sehnal 1989; figure 8.2). The amount of ecdysone involved in molts between larval stages is much less than the amount involved in molts between a larval and a pupal stage. For example, during the course of normal development in the hemimetabolous cockroach, *Nauphoeta cinerea*, ecdysone peaks at 2×0.001 ng ml^{-1} between larval-larval molts, but at 8×0.001 ng ml^{-1} at the penultimate instar (Lanzrein et al. 1985). In the holometabolous insects, larval-to-larval molts (those between instars I to IV in *Bombyx mori*) are initiated by pulses of ecdysone that are roughly two orders of magnitude higher than the baseline titer of ecdysone required for normal tissue development. However, at the larval-to-pupal molt, the titer of ecdysone rises still further, reaching six times that required to initiate the nonmetamorphic larval-larval molt (Nijhout 1994). In fact, during metamorphosis, the level of systemic ecdysone reverts to the very high level present in the embryo prior to hatching. These large quantities of ecdysone initiate the complex reorganizational shifts of metamorphosis in the holometabolous insects. Such large concentrations are not found in hemimetabolous insects nor in spiders (Bonaric and De Reggi 1987).

As part of its metamorphic effects, ecdysone can also act directly on the central nervous system (CNS) of arthropods (Reynolds and Truman 1984; Kovoor 1987; Sakurai et al. 1989; Tautz et al. 1994) to initiate neural sprouting, regression, neuronal remodeling, maturation, and programmed death (Truman 1996). At the outset of metamorphosis, ecdysone receptor proteins allow neuron-by-neuron adjustment (that is, some neurons are retained, others are lost) during cell development (Truman 1996); as metamorphosis progresses, tis-

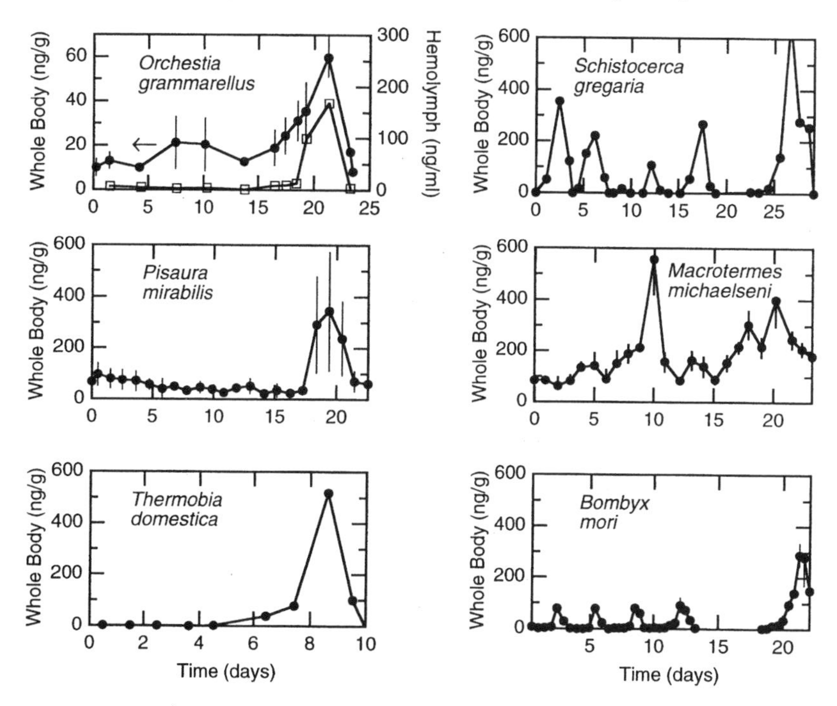

Figure 8.2 Comparative volumes of ecdysone for whole body extract during the development of arthropods. The six graphs compare the amounts of ecdysone (whole animal extracts), during the development of *Orchestia gammarellus,* (reviewed in Spindler 1989), *Pisaura mirablis* (hunting spider; Bonaric and De Reggi, 1987), *Thermobia domestica* (firebrat; Bitsch et al. 1979), *Schistocerca gregaria* (locust; Gande et al. 1979), *Macrotermes michaelseni* (termite; Lanzrein et al. 1985), Bombyx *mori* (silk worm; Prudhomme and Couble 1979; Prudhomme et al. 1985). It may be that the hemolymph volume of ecdysone is a more accurate indicator of hormonal fluctuations attributed to development than whole body extracts because ecdysone affects gene expression even during non-developmental events.

sue response becomes more homogenous as the construction of the adult CNS nears completion (Truman 1996). Finally, ecdysone can act directly on nerve cells to change their electrical activity (Richter et al. 1989), a fact that may be important to silk synthesis in spiders.

Ecdysteroids regulate silk synthesis in the Lepidoptera

Ecdysteroids control silk synthesis in *B. mori* and *Galleria mellonella* by activating a transcriptional hierarchy (Henrich and Brown 1995) (figure 8.3), as well as by stabilizing silk mRNA transcripts (Grzelak et al. 1993). Furthermore, the expression of individual genes coding for the different silk proteins that make up silk fibroins

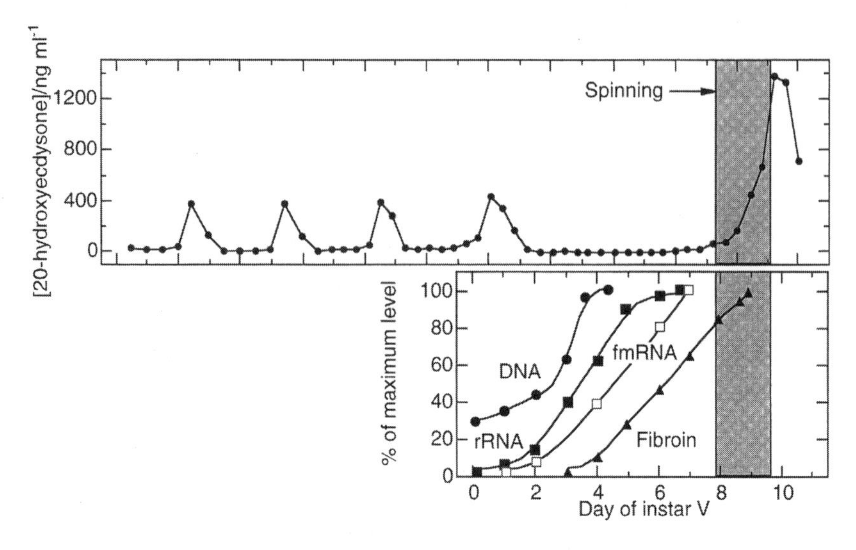

Figure 8.3 Ecdysteroids control ecdysis and silk synthesis by activating a transcriptional hierarchy. (Top) *B. mori* maintains low background concentrations of ecdysone during development that peaks at ecdysis instars 1 to 4. (Bottom) During the fifth instar, which is several days longer than the previous instars, the volume of DNA, rRNA, fmRNA and fibroin increase in sequential order. At Day 8, the amount of ecdysone begins to rise and at about Day 9, silk spinning begins. Very little or no silk is produced in the previous stadia. In contrast, spiders produce silks throughout their development. (After Prudhomme and Couble 1979.)

is not simultaneous (see chapter 2; Grzelak et al. 1993). In *Bombyx mori*, the first four stadia of larval life are characterized by a cyclical pattern of fibroin production. Low background concentrations of ecdysone enable gland development and regulate fibroin transcription (Prudhomme and Couble 1979) (figure 8.3). A general increase in silk produced during each stadia is a function of genome amplification and cell enlargement (Prudhomme et al. 1985); prior to each ecdysis, any silk in the gland is either spun or broken down and reabsorbed (Suzuki et al. 1984). Each larval stadia terminates when a small amount of silk is spun, anchoring the old integument of the animal to its substrate prior to ecdysis (Kiguchi and Agui 1981; Calvez et al. 1976, both cited in Maekawa et al. 1984). None of the silk produced during the first four stadia is retained to contribute to the silks spun at metamorphosis.

Low concentrations of ecdysteroids (5–50 ng ml^{-1}) stimulate RNA synthesis and hence are necessary for silk glands to function (Shigematsu and Moriyama 1970; Kodrík and Sehnal 1994). Larger amounts of ecdysone (5 μg ml^{-1}) exert an inhibitory effect (Kodrík and Sehnal 1994), causing silk gland regression (at larval molts) and degeneration (at pupal molts) (Sehnal and Akai 1990; Townley et al.

1993). Therefore, the amount of ecdysone present must be maintained at a high enough level to allow silk production but not so high that the glands breakdown. It may be that in Lepidoptera, local action of JH regulates the effects of ecdysone on different types of cells. It takes *B. mori* two days to produce enough ecdysone to maximize silk production, and any response decay is tied to the decay of their hormone titer. Regression and degeneration in different silk-producing glands and cells is not simultaneous, but glands and organs disappear at about 48 hours after pupal ecdysis (Sehnal and Akai 1990). From the lepidopteran viewpoint, tying the silk gland's functional response to the hormonal control of insect development appropriately regulates silk production and coordinates it with one period in lepidopteran larval life when massive amounts of silks are needed.

The effects of ecdysone vary spatially and temporally relative to the pattern of its secretion (Ashburner et al. 1974) as well as to the presence and distribution of its receptor proteins. Nuclear hormone receptor genes link extracellular signals directly to gene response (Laudet et al. 1992). Put simply, at the start of metamorphosis, ecdysteroids initiate a complex cascade of sequential gene activation and repression (Jindra and Riddiford 1996). When ecdysone reaches its target cell, it crosses the cell membrane and binds to an intracellular receptor protein, changing the receptor protein's conformation. The intracellular ecdysone-receptor macromolecule then binds with a second receptor protein, a nuclear receptor protein, that acts directly on the gene to activate or inhibit gene expression (Lafont 2000). The specific base sequences to which the receptors bind, the hormone response elements, are scattered throughout the genome (Henrich and Brown 1995). In addition, the products of the first gene transcriptions may activate other genes to produce a delayed secondary response (Henrich and Brown 1995).

JH inhibits the action of ecdysone

The effects of ecdysone are dose-dependent but can be inhibited by the presence of JH (reviewed by Grzelak 1995). Although the molecular mode of action of JH is not well known, it has both developmental and gonadotropic roles. Three primary forms of JH have been identified in insects, but several additional JH-like compounds have also been identified (JH has not been convincingly identified in the Araneae). In insects, JH is secreted by the corpora allata and acts on the brain to regulate the release of PTTH. This, in turn, initiates the synthesis and release of ecdysteroids from the prothoracic gland (Sakurai et al. 1989; Watson et al. 1989; Trabalan et al. 1992; Truman 1996). JH is also produced in the ovaries of adult females

(Riddiford 1996). Most insects secrete only one form of JH, the most commonly secreted form being JH III. The Lepidoptera secrete a mixture of JH-0, JH-I, and JH-II; these differ by the length of their side chains (Gupta 1990; Ross 1991; Nijhout 1994). A new juvenoid, JHB-3, has been recently discovered in the Diptera (Vin 1994), and methyl farnesoate, a precursor of JH, has been identified in the mandibular glands of several decapods (Borst et al. 1987; Laufer et al. 1987). Significantly, although a JH-like compound has been tentatively identified in the Acari (mites and ticks), exogenous application of JH to spiders and other arachnids has had no effect on their development (Kaufman 1997).

Physiologically, JH is difficult to track (Nijhout 1994). Thus, more is known about the effects of JH at the organism level than about how these effects are achieved. Nevertheless, it is known that juvenile hormone is critical in the regulation of gene expression during larval life and early metamorphosis (Riddiford 1985). JH disrupts normal embryonic development and is absent except during the last third of embryogenesis (Riddiford 1994). Early during the last instar of larval development, JH levels are low to undetectable. Ecdysteroid concentrations rise in the absence of JH to initiate metamorphic changes such as the development of flight muscles (Van den Hondel-Franken 1982; Riddiford 1994). In holometabolous insects and when larvae reach a specific size, JH reprograms the brain from a larval to a metamorphic pattern of neurohormonal control (Nijhout 1994). Commitment to a pupal molt is signaled by a declining titer of JH during the last larval instar (Kremen and Nijhout 1989) and an increase in ecdysone that eventually results in metamorphosis (Riddiford 1994).

Although JH acts systemically to control the time of insect metamorphosis, it also acts locally by controlling or influencing the effects of ecdysone on its target tissues (Nijhout 1994). Acting like a steroid molecule, it attaches to specialized binding proteins (JHBP), or it can bind directly with hemolymph proteins to diffuse broadly throughout the organism's body. JH is lipid-soluble and crosses both the cell and nuclear membranes (Nijhout 1994). Hence, its entry into cells is not determined by the presence or absence of specific membrane receptor molecules. Once in the cell, JH is believed to bind with receptor proteins in the cell nucleus. Although the specific molecular mechanisms of JH action are unknown (Kumaran 1990 cited in Grzelak et al. 1993), some investigators propose that JH binds with the one of the same receptor proteins with which ecdysone binds, thereby subverting the ability of ecdysone to initiate or depress protein synthesis (Jones and Sharp 1997). The reconfiguration of JH receptor molecules resulting in the activation of a second messenger system suggests that JH does not act directly on DNA (Nijhout 1994). In contrast, ecdysteroids act directly on the genome and they may also have nongenomic effects (LaFont 2000). At sexual maturity, the prothoracic gland, which is the source of JH synthesis during larval development, degenerates. In many sexually mature female insects, JH is produced in the ovary and has a regulatory function to initiate vitellogenesis or the synthesis of yolk proteins (Nijhout 1994).

Assuming that JH is absent in spiders, it is difficult to imagine why quantitative variation in ecdysone levels and its receptors would not be able to communicate growth and synthetic information, as proposed previously (Emlen and Nijhout 2000). Although nuclear receptor proteins have never been studied in spiders, they are thought to have evolved at least 500 million years ago (Laudet et al. 1992). The nuclear receptor superfamily diverged, primarily through duplication events, and has been well conserved. Furthermore, nuclear receptor genes include large families of genes that code for multiple receptor isoforms and that are composed of several domains (Laudet et al. 1992). Receptor families are likely to perform multiple functions by combinatorial regulation (Henrich and Brown 1995) and cell types vary in their ability to respond to nuclear receptor types. In addition to differentiation by receptor, there are other processes that could affect differential growth, and hence the evolution of castes, in spiders (figure 8.4). One is variation in cell number at the onset of exponential growth. This alone can lead to large differences in adult morphologies in the absence of a secondary, modifying hormone mechanism. Variation in the growth rates of tissues within genotypes

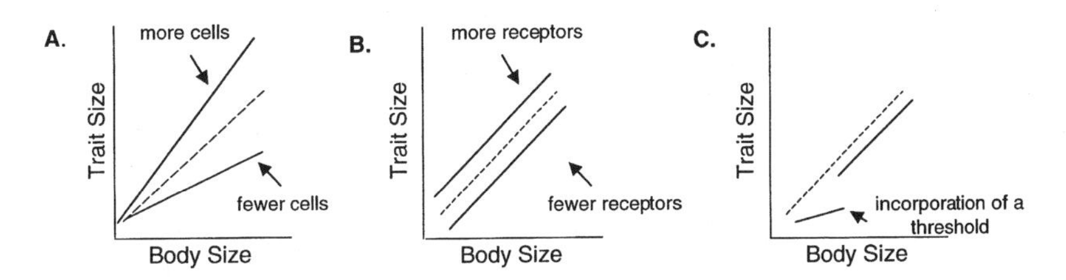

Figure 8.4 Alternative mechanisms for changing scaling relationships among body parts and, hence, the ways in which morphologically differentiated castes might evolve. (A) Changes in the starting conditions, for example, the number of dividing cells present during train growth, may lead to changes in the relative rate of tissue growth and hence exaggerated traits. (B) Changes in the number of receptors for a specific hormone or growth factor may affect how long a tissue grows relative to other body tissue, and hence the relative size of a body part. (C) A regulatory threshold, under the control of a hormonal stimulus, would allow a trait to be expressed only in individuals where a sufficiently high concentration of the hormone was present. As a result, trait expression can jump between minimal to extensive (but size-dependent) to generate a sigmoid or discontinuous scaling of trait relationships. (Redrawn from Emlen and Nijhout 2000 with permission from the *Annual Review of Entomology*, Volume 45. © 2000 by Annual Reviews www.annualreviews.org.)

can also result in different morphologies. The selective activation of gene expression can also have the effect of sudden changes in trait expression (Emlen and Nijhout 2000).

Ecdysone may regulate silk synthesis in some spider glands, but silk production in the *MA* gland seems to be neurally regulated

Regulation of silk synthesis in spiders has not been well studied, and it is not known whether the protein secretions in all silk glands of spiders are under the same mechanism of control. For example, the cylindrical glands of the araneomorph spiders (see chapters 1 and 3) are present only in females and only after sexual maturation. They are the primary source of egg sac silk in most of the Araneomorphae, and silks are assumed to be secreted only when the female spider is producing an egg sac (see chapter 3). This seems to suggest that silk production in the cylindrical glands is tied to egg production and hence could be a response to blood-borne ecdysone.

Protein synthesis in the *MA* silk gland, however, can be induced via direct, mechanical stimulation of silk-producing cells or via the application of cholinergic compounds (Plazaola and Candelas 1991). This suggests a fundamental difference between the regulatory systems for silk synthesis in insects and spiders. A neural signaling system allows precise spatial and temporal control of cell activity. In neural signaling systems, neurotransmitters are released within a few nanometers of the target cell and bind with receptor proteins at the cell surface. Unlike ecdysone and JH, adrenocorticotropic hormone (ACTH) binds with a protein receptor on the cell membrane to trigger a cascade of intracellular processes that ultimately results in silk synthesis. The gland response time, defined by the production of protein by silk-secreting cells, is within 60 minutes and reaches a synthetic peak within 90 minutes (Candelas and Cintron 1981; Plazaola and Candelas 1991). In contrast, transcriptional effects of ecdysteroids take place only after several hours of exposure to the hormone (Lafont 2000). The evolution of neurally driven control of silk synthesis allows spiders to respond much more rapidly to external stimuli than is possible via blood-borne regulators (Peakall 1965; Peakall 1966). Furthermore, dramatic increases in fibroin synthesis can occur at any time during the spider's lifespan. From the spider's viewpoint, neural control of silk synthesis would allow the spider to respond rapidly to the immediate needs posed by its active foraging behavior and a fluctuating environment.

If the silks that spiders use for foraging are neurally controlled, there would seem to be little conflict between silk synthesis and JH as long as silk-producing cells did not contain JH receptor proteins. If this is true, a life-long need to produce silk would not preclude the evolution of eusocial systems.

Developmental flexibility may be a precondition for the evolution of caste systems

Critical size is defined as the size at which the developing larva takes the first physiological steps to metamorphosis (Nijhout 1994). Prior to each molt, insects and spiders may need to grow to a "critical" size that results in ecdysis. The time it takes a larva to reach critical size is frequently determined by nutrition. In general, there is no fixed relationship between the final size of an adult insect and the number of larval molts (reviewed in Emlen and Nijhout 2000). Nevertheless, primitive taxa undergo a larger number of larval molts than do the more derived orders of insects (Nijhout 1994) and spiders (Craig 1987a).

The evolution of the capability to reprogram critical size in response to environmental factors may be the first step to worker polymorphism (Wheeler 1990). The highly skewed morphological differences in worker castes that characterize the leaf cutter ants (Wilson 1980) may indicate one or more developmental reprogramming events. In spiders, the greatest divergence in size is between the sexually dimorphic male and female araneoids in the genera *Nephila* and *Argiope*. Does this indicate that spiders have the developmental capability to evolve morphological castes?

The term "reaction norm" refers to the range of possible morphologies that individuals of the same genotype may attain when reared under different environments or growth conditions. Currently, there are no data available describing the reaction norms or growth patterns of spiders raised under different conditions or spiders with diverse morphologies. The very limited data gathered on the larval growth of the tibia and patella of spiderlings of *A. argentata* raised under identical laboratory conditions suggests that the spiderlings are constrained to reach a specific size before molting to the next stadia (figure 8.5A). The time to reach that size may be variable, (perhaps reflecting spider foraging efficiency; figure 8.5B; S.G. Wolf and C.L. Craig, unpublished). More information is needed to test whether the reaction norm for spider leg size is invariant.

Correlated with spider growth, however, is a shift in spider behavior. Prior to the fourth stadia, spiderlings move off of their regular foraging webs to molt. They then return to the web to resume foraging within an hour or two of molting. After the fourth stadia, however, *A. argentata* females stop spinning webs one to two days prior to ecdysis; during the sixth through the eighth stadia, spinning stops two to three days before ecdysis and does not resume

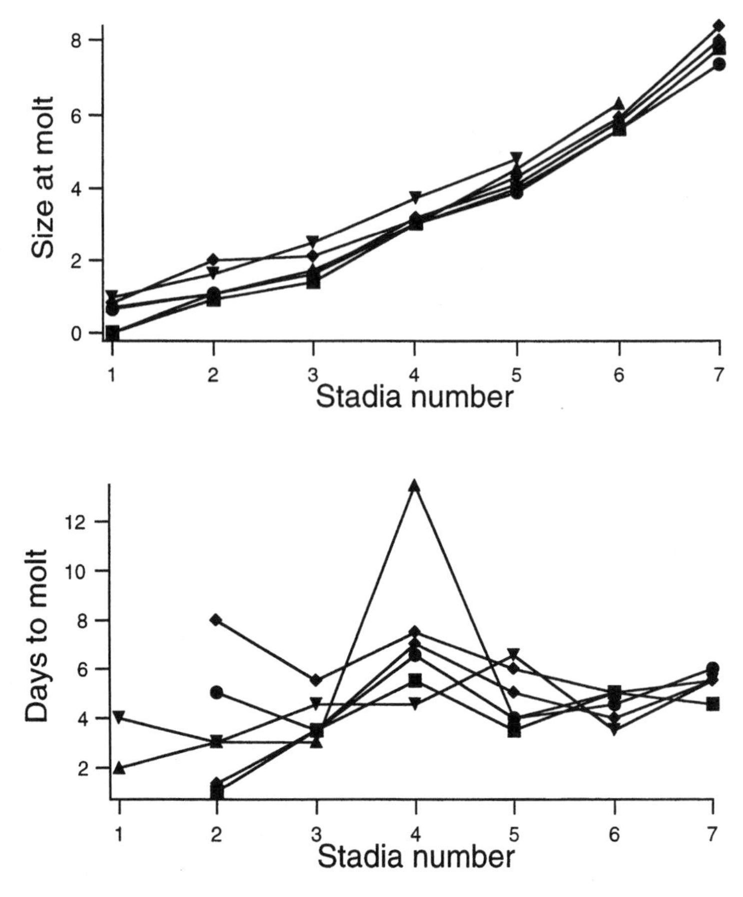

Figure 8.5 Absence of reaction norms for leg length in the developing spider *Argiope argentata*. (Top) Spiders show little variation in size at the end of each stadia or prior to molting. (Bottom) The number of days it takes to reach pre-molting size, however, can be highly variable. (From S.G. Wolf and C.L. Craig, unpublished.)

until one to two days later (pers. observ.). Reduced web production may correlate with a reduction in silk synthesis due to the effects of ecdysone on metamorphosis and hence gland breakdown. In addition, most spiderlings decorate most of the webs they spin during stadia 1 to 4; after the fourth stadia, they spin decorated webs much less frequently (S.G. Wolf and C.L. Craig, unpublished). Thus, it may be that when the spider is small, the effects of decorating behavior to enhance foraging effectiveness are correlated with spider size.

The ant *Solenopsis*, or the fire ant, discussed earlier, shows that the simplest type of developmental switch, a revision of critical size alone, does not lead to the evolution of discontinuous phenotypes or divergent phenotypes (Wheeler 1990). Like these ants, spiders clearly attain critical sizes prior to each ecdysis and in some cases (for example, *Argiope*) may even switch from isometric to anisometric growth. This shift in scaling relationships alone could be achieved via the presence or absence of tissue specific growth receptors (Emlen and Nijhout 2000). The fact that spiders may be able to adjust the length of their intermolt developmental period as well as final adult size in response to food availability suggests that they have the developmental capacity to respond to the environment. However, the difference between the evolution of critical sizes and the evolution of complex morphologies is that in the evolution of complex morphologies some tissues are allowed to grow faster than others. All the evidence so far discussed suggests that spiders have no physiological way of uncoupling the growth rates of different types of tissues.

Conclusion

The silk glands of insects and spiders seem to respond similarly to high concentrations of ecdysone; that is, like any other body tissue, they break down. The presence of JH in Lepidoptera larvae depresses the effects of ecdysone, and specifically silk synthesis. Therefore, with the exception of the tent caterpillars (Fitzgerald 1995), Lepidoptera larvae produce very little silk during larval growth. Silk is not used directly to obtain resources, so there is no obvious cost to the animal for inhibition of silk synthesis.

Spiders present a different picture, however. As in insects, the onset of molting is attained at a specific body size (Nijhout 1994), and growth is inhibited during the ecdysone-controlled molt. However, in araneoid spiders, unlike in insects, silk production is critical during the intermolt period to obtaining the resources needed to reach critical size; if silk production were to be inhibited, then the animal would be unable to grow. Due to the fact that JH is a blood-borne hormone and is lipid-soluble, its effects are general. In insects, JH mediates the resources devoted to silk production (that is, it inhibits silk synthesis) and those devoted to growth. In spiders, the fact that silk synthesis and growth are interdependent would seem to select against the evolution of JH or a JH-like molecule that has such generalized, receptor-independent effects.

If the initiation of silk synthesis is neurally regulated in all spider silk glands, or at least in all but the cylindrical gland, then the spider's life-long need to produce silk, regardless of developmental stage, should not conflict with the regulatory role that ecdysone plays in morphogenesis—silk gland tissue would not respond any differently from any other tissue. On the other hand, even if silk synthesis is neurally regulated, a generalized regulating hormone that acts on the ecdysone molecule could affect silk production. The production of JH receptor proteins would seem to be a relatively simple solution to the silk-and-growth versus no-silk-and-growth conundrum. Whether JH is present or absent, the presence or absence of JH receptor proteins alone could prevent the potential effect of JH on spider silk glands, assuming that JH does not act by binding with ecdysone's receptor proteins, as has been suggested by some authors.

In conclusion, spiders may not have the regulatory hormones, and hence the developmental pathways, that are necessary for the physiological and morphological shifts needed to produce eusocial societies. By some chance evolutionary event, spiders embarked on a developmental pathway that makes eusociality impossible. While the ideas outlined above are, at the least, a plausible solution to the lack of eusociality in spiders, we still do not know if we have all of the pieces of this puzzle, nor how they fit into a final picture of spider social evolution. The next chapter will discuss directions for future research on this problem as well as the other problems posed in previous chapters.

9

Conclusions and Looking Forward

The previous chapters have shown that spiders are defined by their silk systems. From birth to death, the bulk of their behavior depends on silk production. This has apparently been true since their earliest origins and makes spiders unique, different even from other silk-producing animals.

Because of the centrality of silk production to spider lifestyles, we should not be surprised to find that the forces of natural selection have focused on their silk-production systems. As mentioned frequently in previous chapters, the evolution of new silk threads has correlated with major speciation events. Previous chapters discussed the molecular properties of the amino acids that make up silk, the structures of their resulting proteins, and how the genes that encode them are organized. Using these data (collected from both insects and spiders) as background information, silk proteins were studied in the context of the complex and diverse organisms that produce them, looking specifically at the physical properties of silks that are important to function, the mechanisms by which changes in silk proteins are mediated or mitigated, and the energetic demands of silk production.

What makes the study of silk production systems so scientifically rewarding as well as so fascinating is that all aspects of a biological system, from genes to protein structure to function are accessible; silk is easy to collect, and it is easy to observe the ecological effects of different silks in diverse natural habitats. Moreover—and most important for the study of evolution—silk proteins are themselves an ecological phenotype; they are not, as is the case for many other proteins, links in a long chain leading from genotype to phenotype. Due to the unusually short chain from genotype to phenotype represented by silk-production systems, silks provide a uniquely easy way to study the effects of molecular processes on a major behavior, that is, animal foraging. This being so, they also provide a relatively easy way to study the interaction between the physical constraints of a protein structure, the genes that encode it, and selective response for functional properties. If such a complicated interaction can be thoroughly understood in a comparatively simple silk system, the chances are that this understanding will have a major and beneficial impact on the study of the interplay between other molecular processes, natural selection, and macrobehaviors in other organisms.

Natural selection, the repetitive organization of silk genes, and energy exchange are the major factors that direct the evolution of silk proteins and the spiders that produce them

Three major themes have emerged from the results of the research on silk-production systems reviewed in this book: (1) natural selection has been the major factor directing the development of spider silk-producing systems; (2) the highly repetitive organization of silk genes and silk proteins determines the probability of protein innovation, or evolvability; and (3) perhaps most important for the wider study of evolution, silk studies show, for the first time, how the evolution of genes, proteins, and behaviors can be directly linked to quantitative energy costs—specifically, the ATP exchanged between protein synthesis and the breakdown of glucose for all other energy needs that are defined by natural selection.

Studying silk proteins allows a quantitative approach to balancing the effects of genetic innovation, energy economy, and resource optimization in animals

Due to the diverse functional properties of silks, the complex behaviors of insects, and the extraordinarily high cost of secreting silk proteins on a daily basis, we know that natural selection has been the major factor directing the development of spider silk-producing systems. The major selection forces acting on web-spinning spiders have been pressure to produce proteins that can either function in or be easily adapted to new physical habitats, and also pressure to minimize the amount of energy lost to secreted proteins. The results of these forces can be seen in the evolutionary success (measured in speciation events) of spiders that evolve new types of protein threads; the evolutionary success (also measured as speciation events) of spiders that evolve to reconsume and recycle proteins; and the fact that spiders are the most abundant group of organisms that live by predation alone. These phenomena show how genetic innovation, energy economy, and resource optimization can interact in a tangible and quantitative way. Understanding this interaction opens up new perspectives in organismal evolution, for example, why some organisms disperse broadly while others do not.

Silk gene organization enhances protein evolvability while silk protein organization constrains it

We know that the highly repetitive organization of silk genes and silk proteins greatly affects the probability of protein innovation due to genetic effects (for example, mispairing between DNA strands, slippage, duplication, and high cross-over frequency) that act during recombination. Unlike most mutational events, recombination errors affect regions of a gene, not just a single site. Therefore, extensive and rapid genetic effects can be realized in relatively short periods of time, in some cases, over a single generation. The results of the characteristically high probability of gene evolution must allow spiders to adapt to new and diverse environments. Nevertheless, spiders have been around and producing silks for over 350 million years. From the perspective of time, and the comparative evolution of other protein families, one might also ask why it took spiders so long to evolve three types of spun silks. Two of the types of fibroin silks (cocoon silk and *MA* silk) also evolved in other organisms, yet the groups producing them are not as numerous or diverse as the spiders. Despite the innovative effects of the silk gene organization, there must also be a mighty effect that has constrained silk and silk gene evolution.

G.E. Hutchinson gave us the insight that the "evolutionary play" is performed only in an ecological theater—in other words, evolution is context driven. By studying silk genes, expression, and adaptation, we may be able to understand the effects of context on the evolution of one type of protein, fibroin proteins. Nevertheless, the rules affecting the evolution of silk fibroins are not necessarily the same rules that affect the evolution of other types of proteins. Are there common ecological themes that affect protein and organismal evolution alike? Studies of spiders and silks may give insight into the relative weighting of the effects of protein and organismal evolution on organismal speciation.

Balancing ATP invested in silk genes and proteins: a new approach to animal energetics

An earlier report and chapter 8 in this book propose a new means of analyzing animal energetics, here called cost selection. In this approach, the researcher uses ATP as the unit of "currency" when quantifying costs and benefits of various genes, proteins, and metabolic processes. Spiders are the perfect organism to study in this context because they secrete a large and easily measurable volume of protein on a daily basis and because their survival depends on that secretion. Other organisms produce large volumes of secreted proteins such as venom or plant defense compounds that are also an estimatable energy drain on evolving organisms. But the effects of the substitution of one amino acid for another in silks, for example, in comparison to *MA* silk, the substitution of proline in *Flag* silks not only results in a new cost function, but its effect can be directly measured in terms of the organization of the gene, the structure of the protein, and how well the silk intercepts or absorbs the kinetic impact of an intercepted insect. Thus, hypotheses regarding the relationship between natural selection, function, and the energy costs and benefits of gene or protein evolution can be directly tested via manipulative experiments using spiders. Indeed, early studies using this new costing system have already yielded a number of results that are generalizable to other organisms.

Evolutionary analyses of silk protein evolvability, the effects of predators on prey sensory systems, and cost selection are three promising avenues for future evolutionary research and specifically for evolutionary research on spiders

We are still in the early stages of understanding the evolution of silk-producing systems and the animals that make use of them. This work points to three avenues of research that will provide particularly rich insights concerning not just spider evolution but also evolution in general. These three avenues involve examining (1) the relationship between the evolvability of a protein and an organism's adaptation to different habitats; (2) the interplay between the evolution of the capture systems of predators and the sensory, systems of prey; and (3) how resource predictability is related to the predictability of rates of evolutionary change. This last topic is of key importance because it allows study of evolutionary links across multiple spatial and temporal scales. Perhaps most importantly, it allows the ecologist to make new uses of the treasury of molecular genetic and protein data that have become available over the past ten years. It may be possible to relate specific models of the rates that ecologically important proteins evolve to the ability of some organisms to adapt to new physical environments, new prey and new predators using new cost selection approaches.

How is protein evolvability related to adaptation?

Flag silk offers a window into the relationship between the evolvability of a protein and an organism's adaptation to different habitats. The evolution of *Flag* silk correlates with the largest and most rapid speciation event known in spiders. Nevertheless, only one study has characterized the gene in detail. Comparative studies of the genes and proteins across the diverse taxa whose speciation correlates with the evolution of *Flag* could lend new evolutionary insights into the interplay between molecular and organismal "coevolutionary" effects. For example, is the unique organization of the gene the primary factor in spider adaptation, or is it the unique organization and functional properties of the protein that make it so important? Stated more simply, is it the potential for adaptation or the adaptation itself that makes *Flag* silk a keystone innovation for spiders? If such data were available, one could attempt to answer the following questions: How does such variability relate to *Flag*'s gene structure, which is markedly different from other silks' gene structures? Is feedback or regulation involved? How does the rate of change of *Flag* silk compare to the rate of change of *MA* silk? Has *Flag* silk evolved the way it has because it allows greater ecological variability? Or because it reduces error at the genetic level? Answers to these questions might lead to new insights into the pathways of gene selection and evolution, for example, the significance of the presence or absence of introns and the significance of gains or losses during evolution as a whole.

Do predators and prey evolve in tandem?

Spiders' webs offer the perfect arena for examining the interplay between the evolution of prey and the evolution of predator. Nowhere else can this interplay be examined in the field in ways that so little interfere with the animals' natural behaviors. We now know that spiders are evolving in response to insect vision and learning. Are insect visual systems evolving in response to the structure of spider webs? Spiders are, after all, the most abundant predator of insects and affect all insect groups. If spiders and insects are evolving in tandem, it may mean that spiders have had just as great an effect on insect visual systems, flight systems, and behavior as insects have had on the evolution of spiders. In regard to the evolution of predator–prey relationship in general, mapping this interaction could give new depth to our understanding of why insects have evolved in some of the directions that make them the most successful group of organisms (that is, the number of species) on earth. If spiders, their major predator, have not had an effect, why not? Are spiders simply "tracking" insect diversification in response to other selective factors? If this is the case, does the evolution of predation strategies have little or no effect on the evolution of survival strategies of prey? Why are other selective forces—or what are the instances in which other selective forces such as, for example, sexual selection—more important?

Does resource predictability determine rates of evolutionary change?

Once again, because they produce so much protein so constantly, spiders are the natural choice for examining whether selection can act to modify protein structure from one habitat to another. By analyzing silks in a phylogenetic context, in conjunction with using the new ATP approach to energy costing and cost selection, researchers may be able to determine whether there is a way to link resource predictability to rates of evolutionary change. In other words, could researchers determine whether organisms that gain access to extra energy evolve more rapidly? Could the new cost-selection system enable researchers to model these rates of change? If such modeling is possible, it may enable us to link organismal and molecular evolution across the multiple and complex scales at which they occur.

The accessibility of silk systems allows us to link evolutionary effects from gene to organism

It is no coincidence that the first protein conformation characterized was that of a spider silk thread. At a time when analysis methods were cruder than they are today, the relative ease with which spider silk could be collected and examined allowed results that greatly expanded our knowledge of what proteins are and how they function. Today, when analysis methods and our knowledge of genetics, proteomics, and evolution are all still much cruder than we would like, spiders and their silk-production systems still offer us an unparalleled opportunity for discovery. The research advantage is not that silk production is a simpler system than others we might study; it is that the chain of events from genetic organization to macroprocesses such as foraging behavior is shorter. If we come to understand this system, we can make better hypotheses concerning the evolution of different gene and protein expression systems in different ecological settings.

References

Adis, J., and V. Mahnert. 1985. On the natural history and ecology of Pseudoscorpiones (Arachnida) from an Amazonian blackwater inundation forest. *Amazoniana* 9: 297–314.

Alexander, R., K. Noonan, and B. Crespi. 1991. The evolution of eusociality. In *The biology of the naked mole rat* (ed. P. Sherman, J. Jarvis, and R. Alexander), Princeton University Press, Princeton, N.J., pp. 1–44.

Anderson, A. M. 1977. Parameters determining the attractiveness of stripe patterns in the honey bee. *Anim. Behav.* 25: 80–87.

Anderson, J. F. 1990. The size of spider eggs and estimates of their energy content. *J. Arachnol.* 18: 73–78.

Anderson, S. O. 1970. Amino acid composition of spider silks. *Comp. Biochem. Physiol.* 35: 705–711.

Aoki, S. 1977. *Colophina elemstis* (Homoptera, Pemphigidae), an aphid species with soldiers. *Kontyu* 45: 276–282.

Aoki, S. 1982. Soldiers and altruistic dispersal in aphids. In *Biology of Social Insects* (ed. M. Breed, C. Michener, and H. Evans), Westview Press, Boulder, Colo., pp. 154–158.

Arnott, S., S. D. Dover, and A. Elliott. 1967. Structure of β-Poly-L-alanine: refined atomic co-ordinates for an anti-parallel beta-pleated sheet. *J. Mol. Biol.* 30: 201–208.

Ashburner, M., C. Chihara, C. Meltzer, and G. Richards. 1974. On the temporal control of puffing activity in polytene chromosomes. *Cold Spring Harbor Symp. Quant. Biol.* 38: 655–662.

Autrum, H., and Stöcker, M. 1950. Die Verschmelzungsfrequenzen des Bienenauges. *Z. Naturforsch.* 5b, 38–43.

Autrum, H. J., and V. von Zwehl. 1964. Die spektrale Empfindlichkeit einzelner Sehzellen des Bienenaugens. *Z. vergl. Physiol.* 48: 357–384.

Avilés, L. 1992. *Metapopulation biology, levels of selection and sex ratio evolution in social spiders.* Harvard University Press, Cambridge, Mass.

Avilés, L. 1997. Causes and consequences of cooperation and permanent-sociality in spiders. In *Social behavior in insects and arachnids* (ed. J. C. Choe and B. J. Crespi), Cambridge University Press, Cambridge, pp. 476–498.

Babu, K. S. 1973. Histology of the neurosecretory system and heurohaemal organs of the spider, *Argiope aurantia* (Lucas). *J. Morph.* 141: 77–98.

Babu, K. S. 1975. Post embryonic development of the central nervous system of the spider *Argiope aurantia* (Lucas). *J. Morph.* 146: 325–342.

Backhaus, W., and R. Menzel. 1987. Color distance derived from a receptor model of color vision in the honey bee. *Biol. Cybernet.* 55: 321–331.

Barghout, J. Y.L., B.L. Thiel, and C. Viney. 1999. Spider (*Araneus diadematus*) cocoon silk: a case of non-periodic lattice crystals with a twist? *Int. J. Biol. Macromol.* 24: 211–217.

Beckwitt, R., and S. Arcidiacono, 1994. Sequence conservation in the C-terminal region of spider silk proteins (Spidroin) from *Nephila clavipes* (Tetragnathidae) and *Araneus bicentenarius* (Araneidae). *J. Biol. Chem.* 269: 6661–6663.

Beckwitt, R., S. Arcidiacono, and R. Stote. 1998. Evolution of repetitive proteins: spider silks from *Nephila clavipes* (Tetragnathidae) and *Araneus bicentenarius* (Araneidae). *Insect Biochem. Mol. Biol.* 28: 121–130.

Bender, A. 1985. Amino acid metabolism. New York: Wiley.

Berenbaum, M. R., E. S. Green, and A. R. Zangerl. 1993. Web costs and web defense in the parsnip webworm (Lepidoptera: Oecophoridae). *Ann. Am. Entomol. Soc.* 22: 791–795.

Berlese, A. 1913. Intorno alle metamofosi degli insetti. *Redia* 9: 121–136.

Bidmon, H. J. 1988. *Untersuchungen zur Lokalisation von Ecdysteroid-Rezeptoren.* University of Giessen, FRG.

Bitsch, J., A. R. de la Paz, J. Mathelin, J-P Delbecque, J. Delachambre. 1979. *Recherches sur les ecdyséoïdes hémolymphatiques et ovariens de Thermobia domestica* (Insecta Thysanura). *C. R. Acad. Sc.* Paris, 289: 865–868.

Blackledge, T.A., and Pickett, K.M. 2000. Predatory interactions between mud-dauber wasps (Hymenoptera, Sphecidae) and Argiope (Araneae, Araneidae) in captivity. *J. Arachnol.* 28(2): 211–216.

Blackledge, T. A., and J. W. Wenzel. 2000. The evolution of cryptic spider silk: a behavioral test. *Behav. Ecol.* 11: 142–145.

Bonaric, J.-C. 1987. Moulting hormones. In *The ecophysiology of spiders* (ed. W. Nentwig), Springer-Verlag, New York, pp. 111–118.

Bonaric, J. C., and M. De Reggi. 1987. Changes in ecdysone levels in the spider *Pisaura mirabilis* nymphs (Araneae, Pisauridae). *Experientia* 33: 1664–1665.

Bond, J. E., and B. D. Opell. 1998. Testing adaptive radiation and key innovation hypotheses in spiders. *Evolution* 52: 403–414.

Borst, D. W., H. Laufer, M. Landau, E. S. Chang, W. F. Hertz, F. C. Baker, and D. A. Schooley. 1987. Methyl farnesoate and its role in crustacean reproduction and development. *Insect Biochem.* 17: 1123–1127.

Briscoe, A. D., and L. Chittka. 2001. The evolution of color vision in insects. *Annu. Rev. Entomol.* 46: 471–510.

Bristowe, W. S. 1932. The liphistiid spiders with an appendix on their internal anatomy by J. Millot. *Proc. Zool. Soc. London*: 1015–1057.

Brown, K. M. 1981. Foraging ecology and niche partitioning in orb-weaving spiders. *Oecologia* 50: 380–385.

Bruce, M. A., M. E. Herberstein, and M. A. Elgar. 2001. Signalling conflict between prey and predator attraction. *J. Evolution. Biol.* 14: 786–794.

Bückmann, D. 1989. The significance of ecdysone in comparative physiology. In *Ecdysone* (ed. J. Koolman), Thieme Medical Publishers, New York, pp. 20–26.

Burch, C. L., and L. Chao. 2000. Evolvability of an RNA virus is determined by its mutational neighbourhood. *Nature* 406: 625–628.

Burkhardt, D., Darnhofer-Demar, B., and Fischer, K. 1973. Zum binokularen Entfernungsmessung der Insekten. I. Die Struktur des Sehraums von Insekten. *J. Comp. Physiol.* 87: 165–188.

Buskirk, R. 1981. Sociality in the Arachnida. In *Social Insects*, (ed. H. R. Hermann), Volume 4. Academic Press, pp. 282–367.

Calvez, B., M. Hirn, and M. Dereggi. 1976. Ecdysone changes in hemolymph of 2 silkworms (*Bombyx mori*, *Aphilosamia cynthia*) during larval and pupal development. *FEBS Lett.* 71: 57–61.

Campan, R., and M. Lehrer. 2002. Discrimination of closed shapes in two bee species, *Megachile rotundata* and *Apis mellifera*. *J. Exp. Biol.* 205: 559–572.

Candelas, G. C., and J. Cintron. 1981. A spider fibroin and its synthesis. *J. Exp. Zool.* 216: 1–6.

Candelas, G. C., G. Arroyo, C. Carrasco, and R. Dompenciel. 1990. Spider silk glands contain a tissue-specific alanine tRNA that accumulates in vitro in response to the stimulus for silk protein synthesis. *Dev. Biol.* 140: 215–220.

Case, S. 1994. In *Silk polymers* (ed. D. Kaplan, W. W. Adams, B. Farmer, and C. Viney), American Chemical Society, Washington, D.C., p. 544.

Chacon, P., and W. Eberhard. 1980. Factors affecting the numbers and kinds of prey caught in artificial spider webs with consideration of how orb webs trap prey. *Bull. Br. Arachnol. Soc.* 5: 29–38.

Chittka, L. 1992. The color hexagon: a chromaticity diagram based on photoreceptor excitations as a generalized representation of color opponency. *J. Comp. Physiol. A* 170: 533–543.

Chittka, L. 1996. Optimal sets of colour receptors and opponent processes for coding of natural objects in insect vision. *J. Theoret. Biol.* 181: 179–196.

Chittka, L., and R. Menzel. 1992. The evolutionary adaptation of flower colours and the insect pollinators' color vision. *J. Comp. Physiol. A* 171: 171–181.

Coddington, J. 1986. The monophyletic origin of the orb-web. In *Spiders: webs, behavior and evolution* (ed. W. A. Shear), Stanford University Press, Stanford, Calif., pp. 319–363.

Coddington, J. A. 1989. Spinneret silk spigot morphology: evidence for the monophyly of orb weaving spiders, Cyrtophorinae (Araneidae), and the group Theridiidae plus Nesticidae. *J. Arachnol.* 17: 71–95.

Coddington, J. A. 1990. Cladistics and spider classification: Araneomorph phylogeny and the monophyly of orb-weavers (Araneae: Araneomorphae: Orbiculariae). *Acta Zool. Fenn.* 190: 75–97.

Coddington, J. A., and H. W. Levi. 1991. Systematics and evolution of spiders (Araneae). *Annu. Rev. Ecol. Syst.* 22: 565–592.

Coddington, J. A, H. Chanzy, C. L. Jackson, G. Raty, and K. C. H. Gardner. 2002. The unique ribbon morphology of the major ampullate silk of spiders from the genus Loxosceles (recluse spiders). *Biomacromolecules*, 3: 5–8.

Collett, T. S., and L. I. K. Harkness. 1982. Depth vision in animals. In *Analysis of Visual Behavior*, (ed. D. J. Ingle, M. A. Goodale, and R. J. W. Mansfield), M.I.T. Press, Cambridge, Mass., pp. 111–176.

Collett, T. S., and A. Kelber. 1988. The retrieval of visuo-spatial memories by honeybees. *J. Comp. Physiol. A* 163: 145–150.

Couble, P., J. J. Michaille, A. Garel, M. L. Couble, and J. C. Prudhomme. 1987. Developmental switches of sericin mRNA splicing in individual cells of *Bombyx mori* gland. *Dev. Biol.* 124: 431–440.

Coyle, F. A. 1987. The role of silk in prey capture by Nonaraneomorph spiders. In *Spiders: webs, behavior and evolution* (ed. W. A. Shear), Stanford University Press, Stanford, Calif., pp. 267–305.

Craig, C. L. 1986. Orb-web visibility: the influence of insect flight behaviour and visual physiology on the evolution of web designs within the Araneoidea. *Anim. Behav.* 34: 54–68.

Craig, C. L. 1987a. The significance of spider size to the diversification of spider web architectures and spider reproductive modes. *Am. Nat.* 129: 47–68.

Craig, C. L. 1987b. The ecological and evolutionary interdependence between web architecture and web silks spun by orb-web weaving spiders. *Biol. J. Linnean. Soc.* 30: 135–162.

Craig, C. L. 1988. Insect perception of spider orb webs in three light habitats. *Funct. Ecol.* 2: 277–282.

Craig, C. L. 1989. Alternative foraging modes of orb web weaving spiders. *Biotropica* 21: 257–264.

Craig, C. L. 1990. Effects of background pattern on insect perception of spider orb webs. *Anim. Behav.* 39: 135–144.

Craig, C. L. 1994a. Limits to learning: effects of predator pattern and color on perception and avoidance-learning by prey. *Anim. Behav.* 47: 1087–1099.

Craig, C. L. 1994b. Predator foraging behavior in response to perception and learning by its prey: interactions between orb-spinning spiders and stingless bees. *Behav. Ecol. Sociobiol.* 35: 45–53.

Craig, C. L. 1997. The evolution of silks spun by arthropods. *Annu. Rev. Entomol.* 42: 231–267.

Craig, C. L., and G. D. Bernard. 1990. Insect attraction to ultraviolet-reflecting spiders and web decorations. *Ecology* 71: 616–623.

Craig, C. L., and C. F. Freeman. 1991. Effects of predator visibility on prey encounter: a case study on aerial web weaving spiders. *Behav. Ecol. Sociobiol.* 29: 249–254.

Craig, C. L., and R. S. Weber. 1998. Selection costs of amino acid substitutions in ColE1 and Colla gene clusters harbored by *Escherichia coli. Mol. Biol. Evolution* 15: 774-776.

Craig, C. L., A. Okubo, and V. Andreasen. 1985. Effect of spider orb-web and insect oscillations on prey capture. *J. Theoret. Biol.* 115: 201–211.

Craig, C. L., G. D. Bernard, and J. A. Coddington. 1994. Evolutionary shifts in the spectral properties of spider silks. *Evolution* 48: 287–296.

Craig, C. L., R. S. Weber, and G. D. Bernard. 1996. Evolution of predator prey systems: spider foraging plasticity in response to the visual ecology of prey. *Am. Naturalist* 147: 205–229.

Craig, C. L., M. Hsu, D. Kaplan, and N. E. Pierce. 1999. A comparison of the composition of silk proteins produced by spiders and insects. *Int. J. Biol. Macromol.* 24: 109–118.

Craig, C. L., C. Riekel, M. E. Herberstein, R. S. Weber, D. Kaplan, and N. E. Pierce. 2000. Evidence for diet effects on the composition of silk proteins produced by spiders. *Mol. Biol. Evolution* 17: 1904–1913.

Craig, C. L., S. G. Wolf, J. L. Davis, M. E. Hauber, and J. L. Mass. 2001. Signal polymorphism in the web-decorating spider *Argiope argentata* is correlated with reduced survivorship and the presence of stingless bees, its primary prey. *Evolution* 5: 986–993.

Crick, F. H. C., and A. Rich. 1955. Structure of polyglycine II. *Nature* 176: 780–781.

Dabrowska, D., and K. Luczak. 1968. Studies on the incidence of mosquitoes in the food of *Tetragnatha montana* Simon and its food activity in the natural habitat. *Ekol. Polska Ser. A* 16: 843–853.

D'Andrea, M. 1987. Social behaviour in spiders. *Ital. J. Zool.* (N.S.) Monogr. 3.

Dawydoff, C. 1949. *Développement embryonnaire des arachnides.* Traité de Zoologie. P.-P. Grassé. Paris, Masson. 6: 320–395.

Denny, M. W. 1976. The physical properties of spider's silk and their role in the design of orb-webs. *J. Exp. Biol.* 65: 483–506.

Denny, M. W. 1980. Silks - their properties and functions. In *The mechanical properties of biological materials* (ed. J. F. V. Vincent and J. D. Currey), Society for Experimental Biology Symposium XIV. Society for Experimental Biology, New York, pp. 247–271.

Der Ver Sholes, O., and J. Rawlins. 1979. Distribution of orb-weavers (Araneidae: Araneae) in homogeneous old-field vegetation. *Proc. Ent. Soc., Wash.* 81: 234–247.

Dong, Z., R. V. Lewis, and C.R. Middaugh. 1991. Molecular mechanism of spider silk elasticity *Arch. Biochem. Biophys.* 284(1): 53–57.

Dowton, M., and A. D. Austin. 1994. Molecular phylogeny of the insect order Hymenoptera: Apocritan relationships. *Proc. Natl. Acad. Sci. USA* 91: 9911–9915.

Duffey, J. 1996. Eusociality in a coral-reef shrimp. *Nature* 381: 512–514.

Eberhard, W. G. 1986. Effects of orb-web geometry on prey interception and retention. In *Spiders: webs, behavior and evolution,* (ed. W. A. Shear), Stanford University Press, Stanford, Calif., pp. 70–100.

Eberhard, W. G. 1988. Combing and sticky silk attachment behavior by cribellate spiders and its taxonomic implications. *Bull. Br. Arachnol. Soc.* 7: 247–251.

Eberhard, W. G. 1990. Function and phylogeny of spider webs. *Annu. Rev. Ecol. Syst.* 21: 341–372.

Eberhard, W. G., and F. Pereira. 1993. Ultrastructure of cribellate silk of nine species in eight families and possible taxonomic implications (Araneae: Amaurobiidae, Deinopidae, Desidae, Dictynidae, Filistatidae, Hypochilidae, Stiphidiidae, Tengellidae). *J. Arachnol.* 21: 161–174.

Emlen, D. J., and H. F. Nijhout. 2000. The development and evolution of exaggerated morphologies in insects. *Annu. Rev. Entomol.* 45: 661–708.

Enders, F. 1974. Vertical stratification in orb-web spiders (Araneidae: Araneae) and a consideration of other methods of coexistence. *Ecology* 55: 317–328.

Enders, F. 1977. Web-site selection by orb-web spiders, particularly *Argiope aurantia* Lucas. *Anim. Behav.* 25: 694–712.

Endler, J. A. 1988. The color of light in forests and its implications. *Ecological Monographs* 63: 1–27.

Ernst, R., and M. Heisenberg. 1999. The memory template in *Drosophila* pattern vision at the flight simulator. *Vision Res.* 39: 3940–3943.

Espelie, K. E., and D. S. Himmelsbach. 1990. Characterization of pedicel, paper and larval silk from nest of *Polistes annularis* (L.). *J. Chem. Ecol.* 16: 3467–3477.

Espelie, K. E., J. H. Cane, and D. S. Himmelsbach. 1992. Nest cell lining of the solitary bee *Hylaeus bisinuatus* (Hymenoptera: colletidae). *Experientia* 48: 414–416.

Fitzgerald, T. D. 1995. *The tent caterpillars.* Ithaca, Cornell University Press.

Fitzgerald, T. D., and S. C. Peterson. 1988. Cooperative foraging and communication in caterpillars. *Bioscience* 38: 20–25.

Foelix, R. F. 1996. *The biology of spiders.* Oxford University Press, New York.

Franke, S., and G. Käuser. 1989. Occurrence and hormonal role of ecdysteriods in non-arthropods. In *Ecdysone,* (ed. J. Koolman), Thieme Medical Publishers, New York, pp. 294–307.

Frantsevich, L. I., and V. E. Pischka. 1977. Dimensions of the binocular zone of the visual field of insects (English translation from *Zh. Evol. Biokhim. Fiziol.* 1976). *J. Evol. Biochem. Physiol.* 12: 409–412.

Fraser, R. D. B., and T. P. MacRae. 1973. *Conformations of Fibrous Proteins.* New York, Academic Press.

Free, J.B. 1970 Effect of flower shapes and nectar guides on the behaviour of foraging bees. *Behavior* 37: 269–285.

Fujiwara, H., M. Jindra, R. Newitt, S. Palli, K. Hiruma, and L. M. Riddiford. 1995. Cloning of an ecdysone receptor homolog from *Manduca sexta* and the developmental profile of its mRNA in wings. *Insect Biochem. Mol. Biol.* 25: 845–856.

Fukushi, T. 1976. Classical conditioning in the housefly, *Musca domestica. J. Insect Physiol.* 22: 361.

Fukushi, T. 1989. Learning and discrimination of coloured papers in the walking blowfly, *Luclia cuprina. J. Comp. Physiol.* 166: 57–64.

Gabe, M. 1955. Données histologiques sur la neurosécrétion chez les arachnides. *Arch. Anat. Microsc. Morph. Exp.* 44: 351–383.

Gabe, M. 1966. *Neurosecretion.* Pergamon Press, New York.

Gage, L. P., and R. F. Manning. 1980. Internal structure of the silk fibroin gene of *Bombyx mori* II. Remarkable polymorphism of the organization for crystalline and amorphous coding sequences. *J. Biol. Chem.* 225: 9451–9457.

Gande, A. R., E. D. Morgan, and I. D. Wilson. 1979. Ecdysteroid levels throughout the life-cycle of the desert locust, *Schistocerca gregaria*. *J. Insect. Physiol.* 25(8): 669–675.

Garel, J. P., R. L. Garber, and M. A. Siddiqui. 1977. Transfer RNA in posterior silk gland of *Bombyx mori*: polyacrylamide gel mapping of mature transfer RNA, identification and partial structural characterization of major isoacceptor species. *Biochemistry* 16: 3618–3624.

Gatsey, J., C. Hayashi, J. Woods, and R. Lewis. 2001. Extreme diversity, conservation and convergence of spider silk fibroin sequences. *Science* 291: 2603–2605.

Gilmour, D. 1965. The metabolism of insects. W. H. Freeman, San Francisco.

Giribet, G., G. D. Edgecombe, and W. C. Wheeler. 2001. Arthropod phylogeny based on eight molecular loci and morphology. *Nature* 413: 157–161.

Giurfa, M., J. Núñez, L. Chittka, and R. Menzel. 1995. Color preferences of flower-naive honeybees. *J. Comp. Physiol.* A 177: 247–259.

Giurfa, M., B. Eichmann, and R. Menzel. 1996a. Symmetry perception in an insect. *Nature* 382: 458–461.

Giurfa, M., M. Vorobyev, P. Kevan, and R. Menzel. 1996b. Detection of coloured stimuli by honeybees: minimum visual angles and receptor specific stimuli. *J. Comp. Physiol.* 178: 699–709.

Gleadall, I. G., T. Hariyama, and Y. Tsukahara. 1989. The visual pigment chromophores in the retina of insect compound eyes, with special reference to the Coleoptera. *J. Insect Physiol.* 35: 787–795.

Gogala, M. 1967. Die spektrale Empfindlichkeit der Doppelaugen von Ascalaphus macaronius Scop. (Neuroptera, Ascalaphidae). *Z. Vergl. Physiol.* 57: 232–243.

Goldsmith, T. H. 1961. The color vision of insects. In *Light and Life* (ed. W. E. M. Elroy and B. Glass), Johns Hopkins Press, Baltimore, Md., pp. 771–794.

Goldsmith, T. H. 1990. Optimization, constraint, and history in the evolution of eyes. *Q. Rev. Biol.* 65: 281–322.

Goloboff, P. A. 1993. A reanalysis of Mygalomorphae spider families (Araneae). *Am. Mus. Nov.* 3056: 1–27.

Gosline, J. M., and M. E. Demont. 1984. Spider silk as a rubber. *Nature* 309: 551–552.

Gosline, J. M., M. E. Demont, and M. Denny. 1984. The structure and properties of spider silks. *Endeavour* 10: 37–43.

Gosline, J. M., C. C. Pollak, P. A. Guerette, A. Cheng, M. E. Demont, and M. W. Denny. 1994. Elastomeric network models for the frame and viscid silks form the orb web of the spider *Araneus diadematus*. In *Silk polymers* (ed. D. Kaplan, W. W. Adams, B. Farmer, and C. Viney), volume 544. American Chemical Society, Washington, D.C., pp. 328–341.

Gosline, J. M., P. A. Guerette, C. S. Ortlepp, and K. N. Savage. 1999. The mechanical design of spider silks: from fibroin sequence to mechanical function. *J. Exp. Biol.* 202: 3295–3303.

Gould, S. A. C., K. T. Tran, J. C. Spagna, A. M. F. Moore, and J. B. Shulman. 1999. Short and long range order of the morphology of silk from *Latrodectus hesperus* (Black Widow) as characterized by atomic force microscopy. *Int. J. Biol. Macromol.* 24: 151–157.

Greenstone, M. H. 1979. Foraging strategy and metabolic rate in spiders. *Nature* 282: 501–503.

Greenstone, M. H. 1982. Ballooning frequency and habitat predictability in two wolf spider species (Lycosidae: Pardosa). *Florida Entomology* 65: 83–89.

Griswold, C. E., and T. Meikle-Griswold. 1987. *Archaeodictyna ulova,* new species (Araneae: Dictynidae), a remarkable kleptoparasites of group-living eresid spiders. *Am. Mus. Nov.* 2897: 1–11.

Griswold, C. E., J. A. Coddington, G. Hormiga, and N. Scharff. 1998. Phylogeny of the orb-web building spiders (Araneae, Oribuclariae: Deinopoidea, Araneoidea). *Zool. J. Linnean Soc.* 123: 1–99.

Grubb, D. T., and L. W. Jelinski. 1997. Fiber morphology of spider silk: the effects of tensile deformation. *Macromolecules* 30: 2860–2867.

Grzelak, K. 1995. Control of expression of silk protein genes. *Comp. Biochem. Physiol.* 113B: 671–680.

Grzelak, K., B. Kludkiewicz, and Z. Lassota. 1993. The effect of 20-hydroxyecdysone on expression of genes coding for low molecular weight silk proteins of *Galleria mellonella. Insect Biochem. Mol. Biol.* 23: 211–216.

Guerette, P. A., D. G. Ginzinger, B. H. Weber, and J. M. Gosline. 1996. Silk properties determined by gland-specific expression of a spider fibroin gene family. *Science* 272: 112–114.

Gupta, A. P. 1990. Morphogenetic hormones and their glands in arthropods: evolutionary aspects. In *Morphogenetic hormones of arthropods* (ed. A. P. Gupta), Rutgers University Press, New Brunswick, N.J., pp. 3–34.

Hamilton, W. D. 1964. The genetic evolution of social behaviour, I, II. *J. Theoret. Biol.* 7: 1–52.

Hardie, J., and A. D. Lees. 1985. Endocrine control of polymorphism and polyphenism. In *Comprehensive insect physiology biochemistry and pharmacology* (ed. G. A. Kerkut and L. I. Gilbert), Pergamon, New York, pp. 441–490.

Hardie, R. C. 1986. The photoreceptor array of the dipteran retina. *Trends Neurosci.* 9: 419–238.

Hatley, C. L., and J. A. MacMahon. 1980. Spider community organization: seasonal variation and the role of vegetation architecture. *Environ. Ent.* 9: 00–00.

Hauber, M. 1998. Web decorations and alternative foraging tactics of the spider *Argiope appensa. Ethol. Ecol. Evol.* 10: 47–57.

Haupt, J., and J. Kovoor. 1993. Silk-gland system and silk production in Mesothelae (Araneae). *Annales des Sciences Naturelles, Zoologie, Paris* 14: 35–48.

Hausdorf, B. 1999. Molecular phylogeny of araneomorph spiders. *J. Evol. Biol.* 12: 980–985.

Hayashi, C., and R. V. Lewis. 1998. Evidence from flagelliform silk cDNA for the structural basis of elasticity and modular nature of spider silks. *J. Mol. Biol.* 275: 773–778.

Hayashi, C. Y., and R. V. Lewis. 2000. Molecular architecture and the evolution of a modular spider silk protein gene. *Science* 287: 1477–1479.

Henrich, V. C., and N. E. Brown. 1995. Insect nuclear receptors: a developmental and comparative perspective. *Insect Biochem. Mol. Biol.* 25: 881–897.

Hepburn, H. P., and S. P. Kurstjens. 1988. The combs of honeybees as composite materials. *Apidologie* 19: 25–36.

Hepburn, H. R., H. D. Chandler, and M. R. Davidoff. 1979. Extensometric properties of insect fibroins: the green lacewing cross-β, honeybee α-helical and greater waxmoth parallel-conformations. *Insect Biochem.* 9: 69–77.

Herberstein, M. E., C. L. Craig, J. A. Coddington, and M. Elgar. 2000. The functional significance of silk decorations of orb-web spiders: a critical review of their empirical evidence. *Biol. Rev.* 78: 649–669.

Hertz, M. 1930 Die Organisation des optischen Feldes bei der Biene II. *Z. vergl. Physiol.* 11: 107–145.

Higgins, L. E., and R. E. Buskirk. 1992. A trap-building predator exhibits different tactics for different aspects of foraging behavior. *Anim. Behav.* 44: 485–499.

Higgins, L. E., M. A. Townley, E. K. Tillinghast, and M. A. Rankin. 2001. Variation in the chemical composition of orb webs built by the spider *Nephila clavipes* (Araneae, Tetragnathidae). *J. Arachnol.* 29: 82–94.

Hinman, M. B., and R. V. Lewis. 1992. Isolation of a clone encoding a second dragline silk fibroin. *J. Biol. Chem.* 267: 19320–19324.

Holl, A., and M. Henze 1988. Pigmentary constituents of yellow threads of *Nephila* webs. XI Europaisches Arachnologisches Colloquium. Berlin, Technische Universitat Berline (Institut für Biologie FB-14).

Hölldobler, B., and E. O. Wilson. 1990. *The ants.* Harvard University Press, Cambridge, Mass.

Horn, D. H. S. 1989. Historical Introduction. In *Ecdysone* (ed. J. Koolman), Thieme Medical Publishers, New York, pp. 8–19.

Hu, G. G., and W. S. Stark. 1977. Specific receptor input into spectral preference in *Drosophila. J. Comp. Physiol.* 121: 241–252.

Hunt, S. 1970. Amino acid composition of silk from the pseudoscorpion *Neobisium maritumumi* (Leach) a possible link between the silk fibroins and the keratins. *Comp. Biochem. Physiol.* 34: 773–776.

Iisuka, E. 1965. Degree of crystallinity and modulus relationships of silk threads from *Bombyx mori. Biorheology* 3: 1–8.

Ishiwaka, E., and Y. Suzuki. 1985. Tissue and stage-specific expression of sericin genes in the middle silk gland of *Bombyx mori. Dev. Growth Differ.* 27: 73–82.

Jakob, E. M. 1991. Costs and benefits of group living for pholcid spiderlings: loosing food, saving silk. *Anim. Behav.* 41: 711–722.

Jeffreys, A. J., V. Wilson, and S. L. Thein. 1985. Hypervariable 'minisatellite' regions in human DNA. *Nature* 314: 67–73.

Jindra, H., and L. M. Riddiford. 1996. Expression of exdysteriod-regulated transcripts in the silk gland of the wax moth, *Galleria mellonella. Dev. Genes Evol.* 206: 305–314.

Jindra, M., F. Malone, K. Hiruma, and L. M. Riddiford. 1996. Developmental profiles and ecdysteroid regulation of the mRNAs for two ecdysteroid receptor isoforms in the epidermis and wings of the tobacco hornworm, *Manduca sexta*. *Dev. Biol.* 180: 258–272.

Jones, G. J., and P. S. Sharp. 1997. Ultraspiracle: An invertebrate nuclear receptor for juvenile hormones. *Proc. Natl. Acad. Sci. USA* 94: 13499–13503.

Kaiser, W., and E. Liske. 1974. Die optomotorischen Reaktionen von fixiert fliegenden Bienen bei Reizung mit Spektrallichtern. *J. Comp. Physiol.* 80: 391–408.

Kajak, A. 1965. An analysis of food relations between spiders *Araneus cornutis*. *Ekol. Polska Ser. A* 13: 717–761.

Kaufman, W. R. 1997. Subphylum Chelicerata. In *Reproductive biology of invertebrates* (ed. T. S. Adams), Part A. Wiley, New York, pp. 211–245.

Käuser, G. 1989. On the evolution of ecdysteroid hormones. In *Ecdysone* (ed. J. Koolman), Thieme Medical Publishers, New York, pp. 327–336.

Kelber, A. 1996. Colour learning in the hawkmoth *Macroglossum stelatarum*. *J. Exp. Biol.* 199: 1127–1131.

Kelber, A. 1997. Innate preferences for flower features in the hawkmoth *Macroglossum stelatarum*. *J. Exp. Biol.* 200: 827–836.

Kelber, A., and M. Pfaff. 1999. True colour vision in the orchard butterfly. *Naturwissenschaften* 86: 221–224.

Kenchington, W. 1983. The larval silk of *Hypera* spp. (Coleoptera: Cucurlionidae). A new example of cross-β protein conformation in an insect silk. *J. Insect Physiol.* 29: 355–361.

Kent, D., and J. Simpson. 1992. Eusociality in the beetle *Austroplatypus incompertus* (Coleoptera: Curculionidae). *Naturwissenschaften* 79: 86–87.

Kerkam, K., C. Viney, D. Kaplan, and S. Lombardi. 1991. Liquid crystallinity of natural silk secretions. *Nature* 349: 596–598.

Kiritini, K., S. Kawahare, T. Sasaba, and F. Nakasuji. 1972. Quantitative evaluation of predation of spiders on the green leaf hopper, *Neophotettix cincticeps* Uhler, by a sight-count method. *Res. Popul. Ecol.* 13: 187–200.

Kirschfeld, K., and N. Franceschini. 1968. Optische Eigenschaften der Ommatidia im Komplexauge von *Musca*. *Kybernetik* 5: 47–52.

Kirschner, M., and J. Gerhard. 1998. Evolvability. *Proc. Natl. Acad. Sci. USA* 95: 8420–8427.

Kodrík, D. 1992. Small protein components of the cocoons in *Galleria mellonella* (Lepidoptera, Pyralidae) and *Bombyx mori* (Lepidoptera, Bombycidae). *Acta Entomol. Bohemoslov.* 89: 269–273.

Kodrík, D., and F. Sehnal. 1994. Juvenile hormone counteracts the action of ecdysterone on silk glands of *Galleria mellonella* L. (Lepidoptera: Pyralidae). *Int. J. Insect Morphol. Embryol.* 23: 39–56.

Köhler, T., and F. Vollrath. 1995. Thread biomechanics in the two orb-weaving spiders *Araneus diadematus* (Araneae: Araneidae) and *Uloborus walckenaerius* (Araneae, Uloboridae). *J. Exp. Zool.* 271: 1–17.

Komatsu, K. 1985. Chemical and structural studies on silk sericin. Proceedings of the 7th International Wool Textile Research Conferences, Tokyo 1: 373–382.

Kovoor, J. 1987. Comparative structure and histochemistry of silk-producing organs in arachnids. In *The ecophysiology of Spiders* (ed. W. Nentwig and S. Heimer), Springer-Verlag, New York, pp. 160–186.

Krafft, B. 1979. Organisations des societés d'araignées. *J. Psychol.* 1: 23–51.

Kremen, C., and H. F. Nijhout. 1989. Juvenile hormone controls the onset of pupal commitment in the imaginal disks and epidermis of *Precis coenia* (Lepidoptera: Nymphalidae). *J. Insect Physiol.* 35: 603–612.

Kristensen, N. P. 1981. Phylogeny of insect orders. *Annu. Rev. Entomol.* 26: 135–157.

Kuguchi, K., and N. Agui. 1981. Ecdysteroid levels and developmental events during larval molting in the silkworm, *Bombyx mori. J. Insect Physiol.* 27: 805–812.

Kumaran, A. K. 1990. Modes of action of juvenile hormones at cellular and molecular levels. In *Recent Advances in Comparative Arthropod Morphology, Physiology and Development* (ed. A. P. Gupta). Vol. 1, pp. 182–227. Rutgers University Press, New Brunswick.

Kümmerlen, J., J. D. v. Beek, F. Vollrath, and B. H. Meier. 1996. Local structure in spider dragline silk investigated by two-dimensional spin-diffusion nuclear magnetic resonance. *Macromolecules* 29(8): 2920–2928.

Labhart, T. 1974. Behavioral analysis of light intensity discrimination and spectral sensitivity. *J. Comp. Physiol. A* 95: 203–216.

Labhart, T. 1986. The electrophysiology of photoreceptors in different eye regions of the desert ant *Cataglyphis bicolor. J, Comp. Physiol. A*, 158: 1–7.

Labhart, T., B. Hodel, and I. Valenzuela. 1984. The physiology of the cricket compound eye with particular reference to the anatomically specialized dorsal rim area. *J. Comp. Physiol. A* 155: 289–296.

Lafont, R. 2000. Understanding insect endocrine systems: molecular approaches. *Entomol. Exp. Applic.* 97: 123–136.

Lam, S., M. Stahl, K. McMilin, and F. Stahl. 1974. Rec-mediated recombination host spot activity in bacteriophage lambda II: a mutation which causes hot spot activity. *Genetics* 77: 425–433.

Land, M. F. 1989. Variations in the structure and design of compound eyes. In *Facets of vision* (ed. D. G. Stavenga and R. C. Hardie), Springer Verlag, Berlin, Heidelberg, pp. 90–111.

Land, M. F. 1997 Visual acuity in insects. *Annu. Rev. Entomol.* 42: 147–177.

Lanzrein, B., V. Gentinetta, and R. Fehr. 1985. Titres of juvenile hormone and ecdysteroids in reproductive and eggs of *Macrotermes michaelseni*: relation to caste determination? (ed, J. A. L. Watson, B. M. Okot-Kotber, and C. Noirot), 307–327.

Laudet, V., Hänni, J. Coll, F. Catzeflis, and D. Stéhelin. 1992. Evolution of the nuclear receptor gene superfamily. *EMBO J.* 11: 1003–1013.

Laufer, H., D. Borst, F. C. Baker, C. Carrasco, M. Sinkus, D. D. Reuter, L. W. Tsai, and D. A. Schooley. 1987. Identification of a juvenile hormone-like compound in a Crustacean. *Science* 235: 202–205.

Lawrence, J. F., and E. B. Britton. 1991. *Coleoptera. Insects of Australia.* CSIRO. Cornell University Press, Ithaca, N.Y.

Lehrer, M. 1996. Small-scale navigation in the honeybee: active acquisition of visual information about the goal. *J. Exp. Biol.* 199: 253–261.

Lehrer, M. 1997a. Honeybee's visual spatial orientation at the feeding place. In *Orientation and communication in arthropods* (ed. M. Lehrer), Birkhäuser, Basel, pp. 115–134.

Lehrer, M. 1997b. Honeybee's use of spatial parameters for flower discrimination. *Isr. J. Pl. Sci.* 45: 159–169.

Lehrer, M. 1999. Dorsoventral asymmetry of colour discrimination in bees. *J. Comp. Physiol.* 184: 195–206.

Lehrer, M., and S. Bischof. 1995. Detection of model flowers by honeybees: the role of chromatic and achromatic contrast. *Naturwissenschaften* 82: 145–147.

Lehrer, M., and M. V. Srinivasan. 1993. Object-ground discrimination in bees: why do they land on edges? *J. Comp. Physiol. A* 173: 23–32.

Lehrer, M., and M. V. Srinivasan. 1994. Active vision in honeybees: task-oriented suppression of an innate behaviour. *Vision Res.* 34: 511–516.

Lehrer, M., R. Wehner, and V. M. Srinivasan. 1985. Visual scanning behaviour in honeybees. *J. Comp. Physiol. A* 157: 405–415.

Lehrer, M., M. V. Srinivasan, S. W. Zhang, and G. A. Horridge. 1988 Motion cues provide the bee's visual world with a third dimension. *Nature* 332: 356–357.

Lehrer, M., G. A. Horridge, S. W. Zhang, and R. Gadagkar. 1995. Shape vision in bees: innate preference for flower-like patterns. *Philos. Trans. R. Soc. Lond. B* 347: 123–137.

Lewis, A. C., and G. Lipani. 1990. Learning and flower use in butterflies: hypotheses from honey bees. In *Insect–Plant interactions*, (ed. E. A. Bernays), CRC Press, Boca Raton, Fla., pp. 95–110.

Li, S. F. Y., A. J. McGhie, and S. L. Tang. 1994. New Internal structure of spider dragline silk revealed by atomic force microscopy. *Biophys. J.* 66: 1209–1212.

Lin, L. H., D. T. Edmonds, and F. Vollrath. 1995. Structural engineering of an orb-spider's web. *Nature* 373: 146–148.

Lubin, Y. 1973. Web structure and function: the non-adhesive orb-web of Cytophora molluccensis (Doleschall) (Araneae. Araneidae) *Form. Funct.* 4: 337–358.

Lubin, Y. D. 1978. Seasonal abundance and diversity of web-building spiders in relation to habitat structure on Barro Colorado Island, Panama. *J. Arachnol.* 6: 31–51.

Lubin, Y. D. 1986. Web building and prey capture in the Uloboridae. In *Spiders: webs, behavior, and evolution* (ed. W. A. Shear), Stanford University Press, Stanford, Calif., pp. 132–171.

Lubin, Y., W. G. Eberhard, and G. Montgomery. 1978. Webs of *Miagrammopes (*Araneae: Uloboridae) in the neotropics. *Psyche* 85: 580–583.

Lucas, F., and K. M. Rudall. 1967. Variety in composition and structure of silk fibroins: some new types of silk form the Hymenoptera. In

Symposium on fibrous proteins (ed. W. G. Crewther), Plenum Press, New York, pp. 45–54.

Lucas, F., and K. M. Rudall. 1968. Extracellular fibrous proteins: the silks. *Comp. Biochem.* 26B: 475–558.

Lucas, F., J. T. B. Shaw, and S. G. Smith. 1960. Comparative studies of fibroins. *J. Mol. Biol.* 2: 339–349.

Maddison, W. P., and D. R. Maddison. 1992. MacClade: analysis of phylogeny and character evolution. Version 3.0. Sunderland, Massachusetts, Sinauer Associates.

Maekawa, H., N. Saishu-Takada, and N. Miyama. 1984. The effect of 20-hydroxyecdysone on the synthesis of fibroin mRNA in the cultured posterior silk gland of *Bombyx mori*. *Appl. Entomol. Zool.* 19: 341–347.

Magoshi, J., Y. Magoshi, and S. Nakamura. 1994. Mechanism of Fiber Formation of Silkworm. In *Silk polymers* (ed. D. Kaplan, W. W. Adams, B. Farmer, and C. Viney), American Chemical Society, Washington, D.C., pp. 292–310.

Marsh, R. E., R. B. Corey and L. Pauling. 1955a. The structure of Tussah silk fibroin. *Acta Crystallogr.* 8: 710–715.

Marsh, R. E., R. B. Corey, and L. Pauling. 1955b. *Bombyx mori*. *Biochim. Biophys. Acta* 16: 1–34.

Maschwitz, U., K. Dimpert, T. Botz, and W. Rohe. 1991. A silk-nest weaving Dolichoderine ant in a Malayan rain forest. *Insects Socieux* 38: 307–316.

McNamee, S. G., C. K. Ober, L. W. Jelinski, E. Ray, Y. Xia, and D. T. Grubb. 1993. Towards single-fiber diffraction of spider dragline silk from *Nephila clavipes*. In *Silk polymers* (ed. D. Kaplan, W. W. Adams, B. Farmer, and C. Viney), American Chemical Society, Washington, D.C., pp. 168–175.

McReynolds, C. N., and G. Polis. 1987. Ecomorphological factors influencing prey use by two sympatric species of orb-web spiders, *Argiope aurantia* and *Argiope trifasciata* (Araneidae). *J. Arachnol.* 15: 371–383.

Mercer, J. G. 1985. Developmental hormones in parasitic helminthes. *Parsitology Today* 1: 96–100.

Menne, D., and H-C Spatz. 1977. Colour vision in *Drosophila melanogaster*. *J. Comp. Physiol.* 114: 301–312.

Menzel, R. 1967. Intersuchungen zum erlernen von spektralfarben durch die Honiegbien (*Apis melifera*). *Z. vergl. Physiol.* 56: 22–62.

Menzel, R. 1985. Learning in honey bees in an ecological and behavioural context. In *Experimental behavioural ecology* (*Fortschritte der Zoologie 31*), (ed. B. Hölldobler and M. Lindauer), Gustaav Fischer Verlag, New York, pp. 55–74.

Menzel R., and W. Backhaus. 1989. Colour vision in honeybee: phenomena and physiological mechanisms. In *Facets of vision* (ed. D. Stavenga, and R. Hardie), Springer, Berlin, Heidelberg, New York, pp. 281–297.

Menzel, R., and M. Blakers. 1976. Colour receptors in the bee eye – morphology and spectral sensitivity. *J. Comp. Physiol.* 108: 11–33.

Menzel, R., and U. Greggers. 1985. Natural phototaxis and its relationship to colour vision in honeybees. *J. Comp. Physiol. A* 157: 311–321.

Menzel, R., and E. Lieke. 1983. Antagonistic color effects in spatial vision of honeybees. *J. Comp. Physiol.* 151: 441–448.

Menzel, M., J. Erber, and T. Masuhr. 1974. Leaning and memory in the honey bee. In *Experimental Analysis of Insect Behaviour* (ed. L. B. Brown), Spring-Verlag, New York, pp. 195–217.

Menzel, R., D. F. Ventura, H. Herte, J. M. de Souza, and U. Greggers. 1986. Spectral sensitivity of photoreceptors in insect compound eyes: comparison of species and methods. *J. Comp. Physiol. A* 158: 165–177.

Michaille, J.-J., A. Garel, and J.-C. Prudhomme. 1989. The expression of five middle silk gland specific genes is territorially regulated during the larval development of *Bombyx mori. Insect Biochem.* 19: 19–27.

Michaille, J.-J., P. Couble, J. C. Prudhomme, and A. Garel. 1986. A single gene produces multiple sericin messenger RNAs in the silk gland of *Bombyx mori. Biochimie* 68: 1165–1173.

Minet, J. 1994. The Bombycoidea: Phylogeny and higher classification (Lepidoptera: Glossata). *Entomol. Scand.* 25: 63–88.

Mita, K., S. Ichimura, and T. C. James. 1994. Highly repetitive structure and its organization of the silk fibroin gene. *J. Mol. Evol.* 38: 583–592.

Mita, K., S. Ichimura, M. Zama, and T. C. James. 1988. Specific codon usage patterns and its implications on the secondary structure of silk fibroin mRNA. *J. Mol. Biol.* 203: 917–925.

Miyashita, M. 1992. Variability in food consumption rate of natural populations in the spider, *Nephila clavata. Researches on Population Ecology (Kyoto)* 34: 143–153.

Miyashita, T. 1990. Decreased reproductive rate of the spider, *Nephila clavata*, inhabiting small woodlands in urban areas. *Ecol. Res.* 5: 341–351.

Nakamura, T., A. Suyama, and A. Wada. 1991. Two types of linkage between codon usage and gene-expression levels. *FEBS Lett.* 289: 123–125.

Neidhardt, F. C., J. L. Ingraham, and M. Schaechter. 1990. *Physiology of the bacterial cell.* Sinauer Associates, Inc. Sunderland, Mass.

Nelson, M. C. 1971. Classical conditioning in the blowfly (*Phormia regina*): associative and excitatory factors. *J. Comp. Physiol. Psychol.* 77: 353–368.

Nentwig, W. 1983. The non-filter function of orb webs in spiders. *Oecologia* 58: 418–420.

Nentwig, W. 1985. Social spiders catch larger prey: a study of *Anelosimus eximius* (Araneae: Theridiidae). *Behav. Ecol. Sociobiol.* 17: 79–85.

Nentwig, W., and S. Heimer. 1987. Ecological aspects of spider webs. In *Ecophysiology of spiders* (ed. W. Nentwig), Springer-Verlag, New York, pp. 211–225.

Nijhout, F. 1994. *Insect hormones.* Princeton University Press, Princeton.

Nixon, K. C., and J. I. Davis. 1991. Polymorphic taxa, missing values and cladistic analysis. *Cladistics* 7: 233–241.

Noirot, C. 1985. Pathways of caste development in the lower termites. In *Caste differentiation in social insects* (ed. J. A. L. Watson, B. M. Okot-Kotber, and C. Noirot), Pergamon Press, New York, pp. 17–57.

Nyffeler, M., and G. Bentz. 1978. Die Beutespektren der Netzspinnen *Argiope bruennichi* (Scop)., *Araneus quadratus* Cl. und *Agelena labyr-*

inthica (Cl.) in Ödlandwiesen bei Zürich. *Rev. Suisse Zool.* 85: 747–757.

Okubo, A. 1973. A kinetic study of swarming of small animals in motion: Data analysis with *Anarte prichardi* Kim. Chesapeake Bay Institute Technical Report. The Johns Hopkins University, Chesapeake Bay Institute, Baltimore, Md.

Okubo, A., H. E. Chaing, and C. E. Ebbesmeyer. 1977. Acceleration field of individual midges *Anarte pritchardi* (Diptera: Cecidomyiidae), within a swarm. *Can. Entomol.* 109: 149–156.

Olive, C. 1980. Foraging specializations in orb-weaving spiders. *Ecology* 61: 1133–1144.

Opell, B. D. 1985. Web-monitoring forces exerted by orb-web and triangle-web spiders of the family Uloboridae. *Can. J. Zool.* 63: 580–583.

Opell, B. D. 1987. The influence of web monitoring tactics on the tracheal systems of spiders in the family Uloboridae (Arachnida, Araneida). *Zoomorphology* 107: 225–259.

Opell, B. 1990. Material investment and prey capture potential of reduced spider webs. *Behav. Ecol. Sociobiol.* 26: 375–381.

Opell, B. D. 1993. What forces are responsible for the stickiness of spider cribellar threads? *J. Exp. Zool.* 265: 469–476.

Opell, B. D. 1994a. The ability of spider cribellar prey capture thread to hold insects with different surface features. *Funct. Ecol.* 8: 145–150.

Opell, B. D. 1994b. Factors governing the stickiness of cribellar prey capture threads in the spider family Uloboridae. *J. Morph.* 221: 111–119.

Opell, B. D. 1994c. Increased stickiness of prey capture threads accompanying web reduction in the spider family Uloboridae. *Funct. Ecol.* 8: 85–90.

Opell, B. D. 1995. Ontogenetic changes in cribellum spigot number and cribellar prey capture thread stickiness in the spider family Uloboridae. *J. Morph.* 224: 47–56.

Opell, B. D. 1996. Functional similarities of spider webs with diverse architectures. *Am. Naturalist* 148: 630–648.

Opell, B. D. 1997a. A comparison of capture thread and architectural features of Deinopoid and Araneoid orb webs. *J. Arachnol.* 25: 295–306.

Opell, B. D. 1997b. The material cost and stickiness of capture threads and the evolution of orb-weaving spiders. *Biol. J. Linnean Soc.* 62: 443–458.

Opell, B. D. 1998. Economics of spider orb-webs: the benefits of producing adhesive capture thread and recycling silk. *Funct. Ecol.* 12: 613–624.

Opell, B. D. 1999a. Changes in spinning anatomy and thread stickiness associated with the origin of orb-weaving spiders. *Biol. J. Linnean Soc.* 68: 593–612.

Opell, B. D. 1999b. Redesigning spider webs: stickiness, capture area and the evolution of modern orb-webs. *Evol. Ecol. Res.* 1: 503–516.

Opell, B. D. 2001. Cribellum and calamistrum ontogeny in the spider family Uloboridae: linking functionally related but separate silk spinning features. *J. Arachnol.* 29: 220–226.

Opell, B. D., and J. E. Bond. 2000. Capture thread extensibility of orb-weaving spiders: testing punctuated and associative explanations of character evolution. *Biol. J. Linnean Soc.* 70: 107–120.

Opell, B. D., and J. E. Bond. 2001. Changes in the mechanical properties of capture threads and the evolution of modern orb-weaving spiders. *Evol. Ecol. Res.* 3: 567–581.

Palmer, J. 1990. *Comparative morphology of the external silk production apparatus of "primitive" spiders.* PhD Dissertation. Cambridge, Harvard University.

Palmer, J., F. A. Coyle, and F. W. Harrison. 1982. Structure and cytochemistry of silk glands of the mygalomorph spider *Antrodiaetus unicolor* (Araneae, Antrodiaetidae). *J. Morphol.* 174: 269–274.

Parkhe A.D., S. K. Seeley, K. Gardner, L. Thompson, and R. V. Lewis. 1997 Structural studies of spider silk proteins in the fiber. *J. Mol. Recognition* 10: 1–6.

Pasquet, A., and B. Krafft. 1989. Colony distribution of the social spider *Anelosimus eximius. Insects. Soc.* 36: 173–182.

Paulsson, G., U. Lendahl, G. Galli, C. Ericsson, and L. Wieslander. 1990. The Balbiani Ring 3 gene in *Chironomus tentans* has a diverged repetitive structure split by many introns. *J. Mol. Biol.* 221: 331–349.

Paulsson, G., C. Hoog, K. Bernholm, and L. Wieslander. 1992. Balbiani Ring 1 gene in *Chironomus tentans* sequence organization and dynamics of a coding minisatellite. *J. Mol. Biol.* 225: 349–361.

Peakall, D. B. 1965. Regulation of the synthesis of silk fibroins of spiders at the glandular level. *Comp. Biochem. Physiol.* 15: 509–515.

Peakall, D. B. 1966. Regulation of protein production in the silk glands of spiders. *Comp. Biochem. Physiol.* 19: 253–258.

Peakall, D. B. 1971. Conservation of web proteins in the spider *Araneus diadematus. J. Exp. Zool.* 176: 257–264.

Peitsch D., A. Feitz, H. Hertel, J. de Souza, D.F. Ventura, and R. Menzel. 1992. The spectral input systems of hymenopteran insects and their receptor-based colour vision. *J. Comp. Physiol. A* 170: 23–40.

Peters, H. M. 1955. Uber den Spinnapparat von *Nephila madagascariensis. Z. Naturforsch.* 10b: 395.

Peters, H. M. 1982. Wie Spinnen der Familie Uloboridae ihre Beute einspinnen und verzehren. *Verh. Naturwiss. Verh. Hamburg* 25: 147–167.

Peters, H. M. 1983. Struktur und Herstellung der Fangfä den cribellater Spinnen (Arachnida: Araneae). *Verh. Naturwiss. Verh. Hamburg* 26: 241–253.

Peters, H. M. 1984. The spinning apparatus of Ulobridae in relation to the structure and construction of capture threads (Arachnida, Araneida). *Zoomorphology* 104: 96–104.

Peters, H. M. 1986. Fine structure and function of capture threads. In *The ecophysiology of spiders* (ed. W. Nentwig), Springer-Verlag, New York, pp. 187–202.

Peters, H. M. 1992. On the spinning apparatus and structure of the capture threads of *Deinopis subrufus* (Araneae, Deinopidae). *Zoomorphology* 112: 27–37.

Platnick, N. 2002. The world spider catalogue Version 3.0. The American Museum of Natural History. http://research.amnh.org/entomology/spiders/catalog81-87/INTR01.html.

Platnick, N. I., J. A. Coddington, R. R. Forster, and C. E. Griswold. 1991. Spinneret morphology and the phylogeny of Haplogynae spiders (Araneae, Araneomorphae). *Novitates* 3016: 1–311.

Plazaola, A., and G. C. Candelas. 1991. Stimulation of fibroin synthesis elicits ultrastructural modifications in spider silk secretory cells. *Tissue Cell* 23: 277–284.

Pointing, P. J. 1966. A quantitative field study of the predatory behavior by the sheet web spider Frontinella communis on European pine shoot moth adults. *J. Zool.* Canadian 44: 256–273.

Prudhomme, J.-C., and P. Couble. 1979. The adaptation of the silk gland cell to the production of fibroin in *Bombyx mori* L. *Biochimie* 61: 215–227.

Prudhomme, J. C., P. Couble, J. P. Garel, and J. Daillie. 1985. Silk synthesis. In *Comprehensive insect physiology, biochemistry and pharmacology* (ed. G. A. Kerkut and L. I. Gilbert), Pergamon Press, Oxford, pp. 571–594.

Raubenheimer, D., and D. Tucker. 1997. Associative learning by locusts: pairing of visual cues with consumption of protein and carbohydrate. *Anim. Behav.* 54: 1449–1459.

Raven, R. J. 1985. The spider infraorder Mygalomorphae (Araneae): cladistics and systematics. *Bull. Am. Mus. Nat. Hist.* 182: 1–180.

Rees, H. H. 1989. Zooecdysteirods: structures and occurrence. In *Ecdysone* (ed. J. Koolman), Thieme Medical Publishers, New York, pp. 28–38.

Reynolds, S. T., and J. W. Truman. 1984. Eclosion hormone. In *Endocrinology of insects* (ed. G. H. Downer and H. Laufer), Alan R. Liss, New York, pp. 217–233.

Richards, A. M. 1960. Observations on the New Zealand Glow-worm *Arachnocampa luminosa* (Skuse) 1890. *Trans. Royal Soc. New Zealand* 88: 559–574.

Richards, O. W., and R. G. Davies. 1977. *Imms' general textbook of entomology.* Volumes 1 and 2. New York, Chapman & Hall.

Richter, K., Käuser and H.-J. Bidman. 1989. Interaction of ecdysteriods with the nervous system. In *Ecdysone* (ed. J. Koolman), Thieme Medical Publishers, New York, pp. 319–326.

Riddiford, L. M. 1994. Cellular and molecular actions of juvenile hormone I. General considerations and premetamorphic actions. *Adv. Insect Physiol.* 24: 213–274.

Riddiford, L. M. 1996. Juvenile hormone: The status of its "status quo" action. *Arch. Insect Biochem.* 32: 271–286.

Riechert, S. E. 1985. Why do some spider cooperate? *Fla. Entomol.* 68: 105–116.

Riechert, S. E., and J. Luzak. 1982. Spider foraging behavioral responses to prey. In *Spider communication* (ed. P. N. Witt and J. S. Rovner), Princeton University Press, New Jersey, pp. 353–385.

Riekel, C. 2000. New avenues in X-ray microbeam experiments. *Rep. Prog. Phys.* 63: 233–262.

Riekel, C., and F. Vollrath. 2001. Spider silk fibre extrusion: combined wide- and small-angle X-ray microdiffraction experiments. *Int. J. Biol. Macromol.* 29: 203–210.

Riekel, C., C. Branden, C. Craig, C. Ferrero, F. Heidelbach, and M. Müller. 1999a. Aspects of X-ray diffraction on single spider fibers. *Int. J. Biol. Macromol.* 24: 187–195.

Riekel, C., C. L. Craig, D. Burghammer, and S. Müller. 1999b. *Eriophora fuligenea* spider silk: structural studies by scanning X-ray diffraction. *Int. J. Biol. Macromol.* 24: 179–186.

Riekel, C., M. Mueller, and F. Vollrath. 1999c. In situ X-ray diffraction during forced silking of spider silk. *Macromolecules* 32: 4464–4466.

Riekel, C., C. L. Craig, M. Burghammer, and M. Müller. 2001. Microstructural homogeneity of support silk spun by *Eriophora fuliginea* (C.L. Koch) determined by scanning X-ray microdiffraction. *Naturwissenschaften* 88: 67–72.

Robinson, M. H., and B. Robinson. 1970. Prey caught by a sample population of the spider *Argiope argentata* (Araneae, Araneidae) in Panama, a year's census data. *Zool. J. Linnean Soc.* 149: 345–357.

Robinson, M. H., and B. Robinson. 1973. Ecology and behavior of the giant wood spider *Nephila maculata* (Fabricius) in New Guinea. *Smithsonian Contrib. Zool.* 149: 1–74.

Rose, R., and R. Menzel. 1981. Luminance dependence of pigment color discrimination in bees. *J. Comp. Physiol.* 141: 379–388.

Ross, E. S. 1991. Embioptera. *The insects of Australia.* CSIRO. Cornell University Press, Ithaca, N.Y., pp. 405–411.

Rudall, K. M. 1962. Silk and other cocoon proteins. In *Comparative biochemistry* (ed. M. Florkin and H. S. Mason), New York, Academic Press, Volume 4, pp. 397–433.

Rudall, K. M., and W. Kenchington. 1971. Arthropod silks: the problem of fibrous proteins in animal tissues. *Annu. Rev. Entomol.* 16: 73–96.

Ruehlmann, T. W., R. W. Matthews, and J. R. Matthews. 1988. Roles for structural and temporal shelter-changing by fern-feeding lepidopteran larvae. *Oecologia* 75: 228–232.

Rypstra, A. 1982. Building a better insect trap. *Oecologia* 52: 31–36.

Rypstra, A. 1990. Prey capture and feeding efficiency of social and solitary spiders: a comparison. *Acta. Zool. Fenn.* 14: 339–343.

Rypstra, A. L., and R. Tirey. 1989. Observations on the social spider *Anelosimus domingo* (Araneae, Theridiidae), in southwestern Peru. *J. Arachnol.* 17: 368.

Sakurai, S., M. Okuda, and T. Ohtaki. 1989. Juvenile hormone inhibits ecdysone secretion and reponsiveness to prothoracicotropic hormone in prothoracic glands of *Bombyx mori. Gen. Comp. Endocrinol.* 75: 222–230.

Schaller, F. 1971. Indirect sperm transfer by soil arthropods. *Annu. Rev. Entomol.* 16: 407–446.

Scharff, N., and J. A. Coddington. 1997. A phylogenetic analysis of the orb-weaving spider family Araneidae (Arachnida, Araneae). *Zool. J. Linnean Soc.* 120: 355–434.

Schmitz, J., and R. F. A. Morritz. 1998. Molecular phylogeny of Vespidae (Hymenoptera) and the evolution of sociality in wasps. *Mol. Phylogenet. Evol.* 9: 183–191.

Sehnal, F. 1989. Hormonal role of ecdysteriods in insect larvae and during metamorphosis. In *Ecdysone* (ed. J. Koolman), Thieme Medical Publishers, New York, pp. 271–278.

Sehnal, F., and H. Akai. 1990. Insect silk glands: their types, development and function, and effects of environmental factors and morphogenetic hormones on them. *Int. J. Insect Morphol. Embryol.* 19: 79–132.

Seidl, R., and Kaiser, W. 1981. Visual field size, binocular domain and the ommatidial array of the compound eyes in the worker honeybee. *J. Comp. Physiol.* 143, 17–26.

Seki, T., and K. Vogt. 1998. Evolutionary aspects of the diversity of visual pigment chromophores in the class Insecta. *Comp. Biochem. Physiol. B* 119: 53–64.

Seldon, P. 1989. The orb-web weaving spiders in the early Cretaceous. *Nature* 340: 711–713.

Seldon, P., and J. Gall. 1992. A triassic mygalomorph spider from the northern Vosges, France. *Paleontology* 35: 211–235.

Sezutsu, H., and K. Yukuhiro. 2000. Dynamic rearrangement within the *Antheraea pernyi* silk fibroin gene is associated with four types of repetitive units. *J. Mol. Evol.* 51: 329–338.

Sheah, W. K., and D. Li. 2001. Stabilimenta attract unwelcome predator to orb webs. *Proc. R. Soc. Lond. B* 268: 1553–1558.

Shear, W. 1970. The evolution of social phenomena in spiders. *Bull. Br. Arachnol. Soc.* 1: 65–76.

Shigematsu, H., and H. Moriyama. 1970. Effect of ecdysterone on fibroin synthesis on the posterior division of the silk gland of the silkworm, *Bombyx mori*. *J. Insect Physiol.* 16: 2015–2922.

Shultz, J. W. 1987. The origin of the spinning apparatus in spiders. *Biol. Rev.* 62: 89–113.

Simmons, A., E. Ray, and L. W. Jelinski. 1994. Solid-state ^{13}C NMR of *Nephila clavipes*: dragline silk establishes structure and identity of crystalline regions. *Macromolecules* 27: 5235–5237.

Simmons, A. H., C. A. Michal, and L. W. Jelinski. 1996. Molecular orientation and two-component nature of the crystalline fraction of spider dragline silks. *Science* 271: 84–87.

Singer, T. L., K. E. Espelie, and D. S. Himmelsbach. 1992. Ultrastructural and chemical examination of paper and pedicel from laboratory and field nests of the social wasp *Polistes metricus* Say. *J. Chem. Ecol.* 18: 77–86.

Smith, G. B., and J. A. L. Watson. 1991. *Thysanura zygentoma (Silverfish)*. In *The insects of Australia*. CSIRO. Cornell University Press, Ithaca, N.Y., volume 1, pp. 275–278.

Smithers, C. N. 1991. Psocoptera (psocids, booklice). In *The Insects of Australia*. CSIRO. Ithaca, CSIRO, volume 1, pp. 412–420.

Snyder, A. W. 1979. The physics of vision in compound eyes. In *Handbook of sensory physiology* (ed. H. Autrum), Volume VII/6A, Springer, Berlin, Heidelberg, New York, pp. 225–313.

Souza, S. J. D., M. Long, R. J. Klein, S. Roy, S. Lin, and W. Gilbert. 1998. Toward a resolution of the introns early/late debate: only phase zero introns are correlated with the structure of ancient proteins. *Proc. Natl. Acad. Sci. USA* 95: 5094–5099.

Spiller, D. A. 1984. Seasonal reversal of competitive advantage between spider species. *Oecologia* 64: 322–331.

Spindler, K-D. 1989. Hormonal role of ecdysteriods in Crustacea, Chilicerata and other Arthropods. In *Ecdysone* (ed. J. Koolman), Thieme Medical Publishers, New York, pp. 290–295.

Srinivasan, M. V., and M. Lehrer. 1984. Temporal acuity of honeybee vision: behavioral studies using moving stimuli. *J. Comp. Physiol. A* 155: 297–312.

Srinivasan, M. V., and M. Lehrer. 1988. Spatial acuity of honeybee vision and its chromatic properties. *J. Comp. Physiol. A* 162: 159–172.

Srinivasan, M. V., M. Lehrer, S. W. Zhang, and G. A. Horridge. 1989. How honeybees measure their distance from objects of unknown sizes. *J. Comp. Physiol. A* 165: 605–613.

Srinivasan, M. V., M. Lehrer, W. Kirchner, and S. W. Zhang. 1991. Range perception through apparent image speed in freely flying honeybees. *Visual Neurosci.* 6: 519–536.

Stavenga, D. G. 1979 Pseudopupils of compound eyes. In *Vision in invertebrates. Handbook of sensory physiology* (ed. H. Autrum), Volume VII/ 6A. Springer Verlag: Berlin, Heidelberg, pp. 357–439.

Stehr, F. W. 1987. *Immature Insects.* Volumes 1 and 2. Dubuque, Iowa, Kendall/Hunt.

Sterns, S. C. 1989. The evolutionary significance of phenotypic plasticity. *Bioscience* 39: 436–445.

Stryer, L. 1995. *Biochemistry.* W.H. Freeman, New York.

Stubbs, D. G., E. K. Tillinghast, and M. A. Townley. 1992. Fibrous composite structure in a spider silk. *Naturwissenschaften* 79: 231–234.

Suzuki, Y., Y. Tsuda, Y. Tsujimoto, and S. Hirose. 1984. Transcription regulation of silk genes. *Cell Struct. Funct.* 9 (suppl.): S1–S4.

Szent-Gyorgyi, A. G., and C. Cohen. 1957. Role of proline in polypeptide chain configuration of proteins. *Science* 126: 697–698.

Takahashi, Y. 1994. Crystal Structure of silk of *Bombyx mori.* In *Silk Polymers* (ed. D. Kaplan, W. W. Adams, B. Farmer, and C. Viney), Washington, D.C., American Chemical Society, volume 544, pp. 168–175.

Tamura, T., H. Inoue, and Y. Suzuki. 1987. The fibroin genes of *Antherea yamamai* and *Bombyx mori* are different in the core regions but reveal a striking sequences similarity in their 5'-ends and 5'-flanking regions. *Mol. Genet.* 206: 189–195.

Tautz, D., M. Friedrich, and R. Schröder. 1994. Insect embryogenesis – what is ancestral and what is derived? *Development* (Suppl.): 193–199.

Termonia, Y. 1994. Molecular modeling of spider silk elasticity. *Macromolecules* 27: 7378–7381.

Thorne, B. L. 1997. Evolution of eusociality in termites. *Annu. Rev. Ecol. Syst.* 28: 27–54.

Thiel, B., D. Kunkel, and C. Viney. 1994. Physical and chemical microstructure of spider dragline: a study by analytical transmission electron microscopy. *Biopolymers* 34: 1089–1097.

Thiel, B., K. Guess, and C. Viney. 1997. Non-periodic lattice crystals in the hierarchical microstructure of spider (major ampullate) silk. *Biopolymers* 41: 703–719.

Tietjen, W. J. 1986. Effects of colony size on web structure and behavior of the social spider *Mallos gregalis* (Araneae, Dictynidae). *J. Arachnol.* 14: 145–157.

Tillinghast, E. K., E. J. Kavanagh and P. H. Kolbjornsen. 1981. Carbohydrates in the webs of *Argiope* spiders. *J. Morphol.* 169.

Towne, W. F., and W. H. Kirchner. 1989. Hearing in honeybees: detection of air-particle oscillations. *Science* 224: 686–688.

Townley, M. A., E. K. Tillinghast, and N. A. Cherim. 1993. Moult-related changes in ampullate silk gland morphology and usage in the araneid spider *Araneus cavaticus. Philos. Trans. R. Soc. Lond. B* 340: 25–38.

Trabalan, M., A.-M. Bautz, M. Moriniere, and P. Porcheron. 1992. Ovarian development and correlated changes in hemolymphatic ecdysteroid levels in two spiders, *Coelotes terrestris* and *Tegenaria domestica* (Araneae, Agelenidae). *Gen. Comp. Endocrinol.* 88: 128–136.

Truman, J. W. 1996. Steroid receptors and nervous system metamorphosis in insects. *Dev. Neurosci.* 18: 87–101.

Truman, J. W., and L. M. Riddiford. 1999. The origins of insect metamorphosis. *Nature* 40: 447–449.

Tschinkel, W. R. 1988. Colony growth and ontogeny of worker polymorphism in the fire ant, *Solenopsis invicta. Behav. Ecol. Sociobiol.* 22: 103–115.

Tsujimoto, Y., and S. Suzuki. 1979. The DNA sequence of *Bombyx mori* fibroin gene including the 5' flanking, mRNA coding, entire intervening and fibroin protein coding regions. *Cell* 18: 591–600.

Uetz, G. W., A. D. Johnson, and D. W. Schmeske. 1978. Web placement, web structure and prey capture in orb-weaving spiders. *Bull. Br. Arachnol. Soc.* 4: 141–148.

Van den Hondel-Franken, M. A. M. 1982. Critical period of sensitivity to juvenile hormone of the invagination of tracheoblasts into the developing flight muscle fibers of *Locusta migratoria. Gen. Comp. Endocrinol.* 47: 131–138.

Vilhelmsen, L. 1997. The phylogeny of lower Hymenoptera (Insecta), with a summary of the early evolutionary history of the order. *J. Zool. Syst. Evol. Res.* 35: 49–70.

Vin, C. M. 1994. Juvenile hormone III bisepoxide: new member of the insect juvenile hormone family. *Zool. Stud.* 33: 1003–2410.

Voet, D., and J. G. Voet. 1995. *Biochemistry.* New York, John Wiley & Sons.

Vollrath, F. 1982. Colony formation in a social spider. *Zeitschrift fur Tierpsychologie* 60: 313–324.

Vollrath, F., and D. P. Knight. 2001. Liquid crystalline spinning of spider silk. *Nature* 410: 541–548.

Vollrath, F., and E. K. Tillinghast. 1991. Glycoprotein glue beneath a spider web's aqueous coat. *Naturwissenschaften* 78: 557–559.

Vollrath, F., W. J. Fairbrother, R. J. P. Williams, E. K. Tillinghast, D. T. Bernstein, K. S. Gallagher, and M. A. Townley. 1990. Compounds in the droplets of the orb spider's viscid spiral. *Nature* 345: 526–528.

Volny, V. P. and D. M. Gordon 2002. Genetic basis for queen worker dimorphism in a social insect. *Proc. Natl. Acad. Sci.* USA. 99: 6108–6111.

von Frisch, K. 1914. Der Farbensinn und Formensinn der Biene. *Zool. J. Physiol.* 37: 1–238.

von Frisch, K. 1915. Der Farbensinn und Formensinn der Bienen. *Zool. Jb. Abt. all. Zool. Physiol.* 35, 1–182.

von Helversen, O., and W. Edrich. 1974. Der Polarisationsempfänger im Bienen auge: Ein Ultraviolettrezeptor. *J. Comp. Physiol.* 94: 33–47.

Wagner, G. P., and L. Altenberg. 1996. Complex adaptations and the evolution of evolvability. *Evolution* 50: 967–976.

Walker, J. 1987. The amateur scientist: sticky threadlike substances that tend to draw themselves out into bead arrays. *Sci. Am.* 257.

Walker, M. M., and M. E. Bitterman. 1989. Attached magnets impair magnetic field discrimination by honeybees. *J. Exp. Biol.* 141: 447–451.

Warburton, C. 1890. The spinning apparatus of geometric spiders. *Q. J. Micr. Sci.* 3–13.

Warrant, E. J. 2001. The design of compound eyes and the illumination of natural habitats. In *Ecology of sensing* (ed. F. G. Barth, and A. Schmid), Springer Verlag: Berlin, Heidelberg, pp. 187–213.

Warwicker, J. O. 1956. The crystal structure of silk fibroins. *Trans. Faraday Soc.*, 52: 554–570.

Warwicker, J. O. 1960. Comparative studies of Fibroins. II. The crystal structures of various fibroins. *J. Mol. Biol.* 2: 350–362.

Watanabe, T. 1999. Prey-attraction as a possible function of the silk decoration of the uloborid spider *Octonoba sybotides. Behav. Ecol* 5: 607–611.

Watson, J. A. L., and J. J. Sewell. 1981. The origin and evolution of caste systems in termites. *Sociobiology* 6: 101–118.

Watson, R. D., E. Spaziani, and W. E. Bollenbacker. 1989. Regulation of ecdysone biosynthesis in insects and crustaceans: a comparison. In *Ecdysone* (ed. J. Koolman), Thieme Medical Publishers, New York, pp. 188–220.

Wehner, R. 1981. Spatial vision in arthropods. Invertebrate visual centers and behavior. In *Handbook of sensory physiology* (ed. H. Autrum), volume VII/6C. Springer-Verlag, Berlin, pp. 287–616.

Wehner, R., and S. Strasser. 1985. The POL area of the honey bee's eye: behavioural evidence. *Physiol. Entomol.* 10: 337–349.

Weiss, M. 1995. Associative learning in a nymphalid butterfly. *Ecol. Entomol.* 20: 298–301.

West-Eberhard, M. J. 1989. Phenotypic plasticity and the origins of diversity. *Annu. Rev. Ecol. Syst.* 20: 249–278.

Wheeler, D. E. 1986. Developmental and physiological determinants of caste in social Hymenoptera: evolutionary implications. *Am. Naturalist* 128: 13–34.

Wheeler, D. E. 1990. The developmental basis of worker polymorphism in fire ants. *J. Insect Physiol.* 36: 315–322.

Wheeler, D. 1991. The developmental basis of worker caste polymorphism in ants. *Am. Naturalist* 138: 1218–1238.

Wheeler, D. E., and H. F. Nijhout. 1984. Soldier determination in *Pheidole bicarniata:* inhibition by adult soldiers. *K. Insect Physiol.* 30: 127–135.

Wiehle, H. 1931. Neue Beiträge zur Kenntnis des Fanggewebes der Spinnen aus den Familien Argiopidae Uloboridae und Theridiidae. *Z. Morphol. Ökol* 22; 349–400.

Wilson, E. O. 1953. The origin and evolution of polymorphism in ants. *Q. Rev. Biol.* 28: 136–156.

Wilson, E. O. 1971. *The insect societies.* Harvard University Press, Cambridge, Mass.

Wilson, E. O. 1980. Caste and division of labor in leaf-cutter ants. *Ecol. Sociobiol.* 7: 143–156.

Wise, D. H. 1981. Inter- and intraspecific effects of density manipulations upon females of two orb-weaving spiders (Araneae: Araneidae). *Oecologia* 48: 252–256.

Wise, D. H., and J. L. Barata. 1983. Prey of two syntopic spiders with different web structures. *J. Arachnol.* 11: 271–281.

Wolf, E. 1934. Das Verhalten der Biene gegen, ber flimmernden Feldern und bewegten Objekten. *Z vergl Physiol* 20: 151–161.

Work, R. W., and N. Morosoff. 1982. A physico-chemical study of the supercontraction of spider major ampullate silk fibers. *Textile Res. J.* 52: 349–356.

Xu, M., and R. V. Lewis. 1990. Structure of a protein superfiber: spider dragline silk. *Proc. Natl. Acad. Sci.* USA 87: 7120–7124.

Yang, Z., D. Grubb, and L. Jelinski. 1997. Spider major ampullate silk (drag line): smart composite processing based on imperfect crystals. *Macromolecules*: 8254. 30: 8254–8261

Zhou, C.-Z., F. Confalonieri, N. Medina, Y. Zivanovic, C. Esnault, T. Yang, M. Jacquet, J. Janin, M. Duguet, et al. 2000. Fine organization of *Bombyx mori* fibroin heavy chain gene. *Nucleic Acids Res.* 28: 2413–2419.

Index